Biotechnology

A Laboratory Course

Biotechnology ———————

A Laboratory Course

Jeffrey M. Becker

Department of Microbiology
Graduate Program in Cellular, Molecular, and
    Developmental Biology
The University of Tennessee
Knoxville, Tennessee

Guy A. Caldwell

Graduate Program in Cellular, Molecular, and
    Developmental Biology
The University of Tennessee
Knoxville, Tennessee

Eve Ann Zachgo

Microbial Products and Processes Department
BioTechnica International
Cambridge, Massachusetts

Academic Press, Inc.
Harcourt Brace Jovanovich, Publishers
San Diego   New York   Berkeley   Boston   London   Sydney   Tokyo   Toronto

Copyright © 1990 by Academic Press, Inc.
All Rights Reserved.
No part of this publication may be reproduced or transmitted in any form or by any
means, electronic or mechanical, including photocopy, recording, or any information
storage and retrieval system, without permission in writing from the publisher.

Academic Press, Inc.
San Diego, California 92101

United Kingdom Edition published by
Academic Press Limited
24–28 Oval Road, London NW1 7DX

Library of Congress Cataloging-in-Publication Data

Becker, Jeffrey M.
    Biotechnology : a laboratory course / Jeffrey M. Becker, GuyA.
  Caldwell, Eve Ann Zachgo.
      p. cm.
    ISBN 0-12-084560-1 (pbk. : alk. paper)
    1. Biotechnology--Laboratory manuals.   I. Caldwell, Guy A.
  II. Zachgo, Eve Ann. III. Title.
    [DNLM: 1. Biotechnology--laboratory manuals.   TP 248.24 B395b]
  TP248.2.B374   1990
  660'.6--dc20
  DNLM/DLC
for Library of Congress   89-17915
                                                           CIP

Printed in the United States of America
90 91 92 93   9 8 7 6 5 4 3 2 1

To the parents of Jeff Becker,
        Dorothy and Harry Becker

To the parents of Guy Caldwell,
        Marc and Victoria Caldwell

To the parents of Eve Zachgo,
        Frank and Corinne Zachgo
and to her sister,
        Emily Zachgo

Contents _____

Introductory Notes
Record Keeping and Safety Rules

Exercise 1

Exercise 2

Exercise 3

Preface

This book is different. To our knowledge, there is no text designed for instruction at the introductory or even advanced level that is comparable to this manual. We have attempted to create a text that consists of a series of laboratory exercises, beginning with basic techniques and culminating in the utilization of previously acquired technical experience and experimental material. Thus, this manual provides a continuum of experiments using the same biological material throughout, which is beneficial in understanding a common sequence of technical stages used in biotechnology in a framework which is suited to instructional methodologies.

This manual is an outgrowth of an introductory laboratory course for senior undergraduate and first year graduate students in the biological sciences at The University of Tennessee, which was designed to provide students with an in-depth experience and understanding of selected methods, techniques, and instrumentation used in modern biotechnology. A continuum was established by using two principal organisms: the bacterium *Escherichia coli* and the yeast *Saccharomyces cerevisiae*. In addition, a single plasmid (pRY121) and a single enzyme (β-galactosidase) for demonstrating various biotechnological manipulations were utilized. Because our main objective was to provide a step-by-step laboratory course for students, there are aspects of molecular biology and biotechnology which are not included.

To facilitate the use of this manual by diverse undergraduate institutions, we have provided a suggested weekly schedule based on a fifteen-week format, which we have found useful at The University of Tennessee; ensured ready availability of the few microbial cultures (*Escherichia coli* and *Saccharomyces cerevisiae*) necessary for this course; attempted to design the course for a minimum number of reagents and supplies; and do not propose work with radioactivity.

The continued impact of recombinant DNA technology in both academic and industrial research has resulted in a substantial change in the curriculum of graduate and undergraduate biology programs. This increased emphasis on new technology required a novel and contemporary method of instruction. It is our hope that this manual will aid academic institutions in their efforts to establish a laboratory course in molecular biology and biotechnology.

Jeffrey M. Becker
Guy A. Caldwell
Eve Ann Zachgo

Acknowledgments ———————————————————

We would like to acknowledge the following individuals and organizations for their time, effort, and support in the preparation of this manual: Paula Keaton and Marty Danilchuk for their dedicated service in typing the manuscript; Kim Neifer, Wendy Schilling, and Kumar Alagramam for their useful comments and advice on the experimental protocols; Gary Sayler, Frank Larimer, Don Dougall, Chuck Pettigrew, Min Park, Robert Kreisberg, and Pat Rayman for their cumulative efforts in the instruction and design of the course; R. Rogers Yocum for use of plasmid pRY121; David Miller for photography services; and the Science Alliance, State of Tennessee Center for Excellence in the Life Sciences, for the laboratory facilities and funding of the biotechnology laboratory at The University of Tennessee, Knoxville.

Suggested Schedule for Exercises _____

Week	Day 1	Day 2	Day 3	Day 4
1	Introduction/ Orientation		Exercise 1 Exercise 2	
2	Exercise 3 Exercise 4		Exercise 5	
3	Exercise 6*		Exercise 7	
4	Exercise 8, Parts A and B		Exercise 8, Part C	
5	Exercise 9,† Day 1	Exercise 9, Day 2		Exercise 9, Day 4
6	Exercise 9, Day 7		Exercise 10	
7	Exercise 11, Day 1	Exercise 11, Day 2	Exercise 11, Day 3	

* Exercise 6A (Isolation and Characterization of Auxotrophic Yeast Mutants) is an additional exercise. (See Appendix 1.)

† Exercise 9A (Large-Scale Isolation of Plasmid DNA by Column Chromatography) is an alternative procedure which takes 2 days. (See Appendix 1.)

Week	Day 1	Day 2	Day 3	Day 4
8	Exercise 12, Day 1		Exercise 12, Day 2	
9	Exercise 12, Day 3		Exercise 12, Day 4	Exercise 12,* Day 5
10	Exercise 13		Exercise 14, Day 1	Exercise 14, Day 2
11	Exercise 15		Exercise 16	
12	Exercise 17		Exercise 18	
13	Exercise 19, Part A, Day 1		Exercise 19, Part A, Day 2	
14	Exercise 19, Part B, Day 1	Exercise 19, Part B, Day 2	Exercise 19, Part B, Day 3	
15	Exercise 19, Part C, Day 1	Exercise 19, Part C, Day 2		

* Exercise 12A (Colony Hybridization) is an additional exercise which takes 2 days more. (See Appendix 1.)

Record Keeping and Safety Rules

Format of Student Laboratory Records

Ten Dos and Don'ts of Record Keeping

Criteria for Grading the Laboratory Notebook

Safety Rules in the Laboratory

Format of Student Laboratory Records

These guidelines should be followed for the correct recording of the exercises.

A. Purpose/Introduction

Two or three sentences stating the exact purpose(s) and objective(s) of the exercise; what was done, to what, and why. Do not include, as a part of this abstract, familiarization with equipment or techniques unless applicable. A *short* conclusion giving results/ideas may be added, if applicable.

B. Materials and Methods

Include, *only,* any deviations or additional equipment/procedures from those stated in the laboratory manual. If none, merely note "No deviations from the planned procedure were performed." If deviations were minor they may be ignored. However, accurate and complete cross-referencing to any references or procedures is mandatory.

C. Data/Results

All data generated should be recorded *during* the exercise. This includes recopying any tables, graphs, formulas, etc., from the laboratory manual and/or drafting new tables, graphs (labeled), as necessary, to accurately and *neatly* represent the data. This section should also include all calculations, averages, error analyses, and corrections of recorded data.

D. Discussion and Conclusions

This *should* include any interpretations, conclusions, or suggestions regarding the results obtained; perhaps a discussion of expected results and why they were or were not achieved, giving evidence for your views, including any assumptions you find it necessary to make. THIS IS *NOT* TO BE AN ENTIRE SUMMARY OF THE EXPERIMENT—ONLY A DISCUSSION OF THE FINAL RESULTS!

E. References

Include any that were consulted for the experiment or made in reference to the report. This is usually *at least* your laboratory manual.

Some Important Notes
1. *At all times* be concise and honest.
2. Label *all* tables and graphs, indicating what they are supposed to represent.
3. DO NOT use different color pens or pencils in any portion of the reports.
4. Graphs should be numbered on axes in easy to work with divisions (i.e., 5, 10, 15, 20, etc., not 2.3, 4.6, 6.9, etc.).
5. *Remember,* this is *not* a personal diary and should not include any such references.
6. *Remember,* you are *not* being graded on your results; therefore, a good, accurate representation of them, with any intelligent explanation or hypothesis for any data obtained, is critical.
7. If available, the use of computers in conjunction with word processing software and spreadsheets may facilitate report writing and data analysis.

Ten Dos and Don'ts of Record Keeping*

1. **Do** keep the record factual.
   Don't editorialize.

2. **Do** use a record book with a permanent binding.
   Don't use a loose-leaf, spiralbound, or any other type of temporarily bound book that allows page deletions and insertions.

3. **Do** make two copies of all notebook entries, one of which should be kept safely at a separate location.
   Don't have only one copy of notebook susceptible to loss or damage.

4. **Do** enter data and information directly into the record book promptly as generated. You may wish to sign and date each page of the record book at the time of entry (signing is required procedure for industrial research notebooks).
   Don't rely on memory or use informal loose sheets for entries with the intention of later putting them into the bound record book.

5. **Do** use a permanent ink, preferably black, which will reproduce well when photocopied.
   Don't use a pencil or nonpermanent or colored inks.

6. **Do** identify errors and mistakes and explain them.
   Don't ignore errors and mistakes or obliterate, delete, or otherwise render them unreadable.

*Derived from "Record Books: Their Generation, Maintenance, and Safekeeping." Dow Chemical Company, Midland, Michigan.

7. **Do** attach support records to the record book or store such records, after properly referencing and cross-indexing, in a readily retrievable manner.
   Don't file supporting records in a haphazard manner without any record of their relationship or connection to the research reported in the record book.

8. **Do** use standard accepted terms; avoid abbreviations, code names, or numbers if possible.
   Don't use any abbreviation, code name, or number without giving its meaning or definition.

9. **Do** keep the record book clean; avoid spills and stains.
   Don't subject the pages of the research notebook to chemical or physical destruction from spills.

10. **Do** keep a table of contents, and index the record book as soon as it is filled.
    Don't leave notebook entries without proper filing or indexing.

Criteria for Grading the Laboratory Notebook

Criterion	Possible points
1. Factuality of record (avoidance of editorializing)	5
2. Use of permanent ink	5
3. Legibility/Neatness	10
4. Proper correction and explanation of errors	5
5. Attachment of support records	5
6. Use of standard, accepted terms and defined abbreviations	10
7. Proper cross-referencing to techniques, results, and support records	15
8. Format	
A. Purpose/Introduction	8
B. Materials and Methods	8
C. Data/Results	8
D. Discussion and Conclusions	8
E. References	8
9. Table of Contents	5
	100

*Instructor's Note*_____

These suggested criteria have been found quite useful in evaluation of student notebooks. You may wish to use these guidelines as a basis for grading written laboratory reports. The combined use of the Format of Student Laboratory Records, the Ten Dos and Don'ts of Record Keeping, and the Criteria for Grading the Laboratory Notebook provides the student with a concrete format and defined expectations to follow in recording laboratory work.

Safety Rules in the Laboratory

1. Prepare for each laboratory period by reading each exercise and by becoming familiar with the principles and methods involved. Familiarity with the exercise decreases your chances of an accident. Also, advance preparation allows you to use your time efficiently to complete the experiment.

2. No eating, drinking, or smoking is permitted in the laboratory.

3. Lab coats or aprons must be worn at all times in the laboratory. This helps ensure that no culture material is accidentally deposited on your clothes or skin, and protects you and your clothes from spillage of chemicals and stains.

4. Only those materials pertinent to your lab work, such as lab manuals, lab notebooks, and other lab materials, should be brought to your laboratory work space. All other items, such as coats, books, and bags, should be stored away from your work area.

5. Begin each laboratory session by disinfecting your work area. Saturate the area with a disinfectant, spread the disinfectant with a paper towel, and allow the area to dry. Repeat this procedure after you have finished your work to ensure that any material you have deposited on the work surface is properly disinfected.

6. All culture material and chemicals should be properly labeled with your name, class, date, and experiment. Labeling is critical to avoid improper use or disposal of material.

7. Be very careful with Bunsen burners. To avoid injuries, burners should be turned off when not in use. When reaching for objects, be careful not to place your hands into the flame.

8. All contaminated material must be disinfected before disposal or reuse. All material to be autoclaved should be placed in a proper receptacle for collection. Used pipets should be placed in disinfectant.

9. After the laboratory session, observe good hygiene by washing your hands before leaving the laboratory.

10. In the event of any accident or injury, report immediately to the laboratory instructor so that prompt and proper action can be taken.

1

Measurement of pH

Introduction

Maintenance of pH with the use of inorganic or organic buffering systems is a necessary part of any protocol designed for the study of biological systems. This laboratory exercise will demonstrate the variation in pH that can be obtained with an inorganic buffer system and how pH is measured both electronically with a pH meter and visually with a pH-sensitive dye.

pH is measured in a number of different ways. An accurate and practical method for measuring pH involves the use of a pH meter. The pH meter is a potentiometer which measures the potential developed between a glass electrode and a reference electrode. In modern instruments, the two electrodes are frequently combined into one electrode, known as a combination electrode.

The glass electrode contains a glass bulb constructed of very thin, special glass that is permeable to hydrogen ions. As a result, a potential develops across the glass membrane. The potential is linearly related to the pH. Standardization against a buffer of known H^+ concentration is required since the concentration of H^+ inside the bulb of the glass electrode changes with time (see Appendixes 2, 3, and 4, Buffer Solutions, Preparation of Buffers and Solutions, and Properties of Some Common Concentrated Acids and Bases, respectively). Adjustments for temperature are necessary since the relationship between measured potential and pH is temperature dependent. In actuality, there are a number of other potentials which develop in this system (e.g., liquid junction potential and asymmetry potential), but they are usually constant and relatively independent of pH so that the linearity of measured potential with pH is maintained.

Instructor's Note

This exercise may be dispensable depending on the level and experience of the students. However, we have found this exercise useful for all students in the orientation and familiarization with the laboratory environment.

Reagents/Supplies

Beakers (250 ml)
Bromthymol blue [0.25% (w/v)]
pH indicator paper
Pipets (1, 5, and 10 ml)
Potassium phosphate, dibasic (K_2HPO_4)
Potassium phosphate, monobasic (KH_2PO_4)
Standard buffer (pH 4.01)
Standard buffer (pH 7.00)
Test tubes (18 × 150 mm)
Wash bottle

Equipment

pH meter

Instructor's Note

For the novice student, reading of Appendixes 2 and 3 for appropriate background material is essential for understanding buffer preparation.

Procedure

Part A. Visual Estimation of pH

1. Make up 0.1 *M* solutions (100 ml) of K_2HPO_4 and KH_2PO_4.

2. Set up a series of 12 test tubes (18 × 150 mm) as shown in Table 1.1.

3. To tubes 1, 3, 5, 7, 9, and 11, add 5 drops of bromthymol blue; mix. You now have a series of color standards covering the pH range 5.30–7.73. Record your observations. This exercise simply illustrates that indicator dyes may be useful as pH indicators.

Part B. Measurement of pH with a pH Meter

1. Standardize the pH meter using the standard pH 7.00 buffer in a 250-ml beaker. Rinse the electrodes using a wash bottle. Do not wipe the electrodes with tissue because this creates a static electric

Table 1.1   Phosphate Buffers

Tube	0.1 *M* K_2HPO_4 (ml)	0.1 *M* KH_2PO_4 (ml)	pH expected
1	0.3	9.7	5.30
2	0.5	9.5	5.59
3	1.0	9.0	5.91
4	2.0	8.0	6.24
5	3.0	7.0	6.47
6	4.0	6.0	6.64
7	5.0	5.0	6.81
8	6.0	4.0	6.98
9	7.0	3.0	7.17
10	8.0	2.0	7.38
11	9.0	1.0	7.73
12	9.5	0.5	8.04

charge on the electrodes and may cause erroneous readings. Remove the last drop of water by carefully touching a piece of clean tissue paper to the drop.

2. Measure the pH of the standard pH 4.01 buffer. Reset the pH meter if necessary. It is important to measure the pH with two standard buffers to ensure that the pH meter is functioning properly over the entire pH range.

3. Measure the pH of the solutions in tubes 2, 4, 6, 8, 10, and 12 (prepared in Part A) with the pH meter. Rinse the electrodes between readings and handle them carefully. You may also wish to use pH indicator paper to get an idea of the pH of the solutions. This rapid method is often accurate enough for some applications and is especially useful for very small volumes or radioactive solutions.

4. Record your observations from Parts A and B. Correlate the measured values from Part B to the expected pH values from Table 1.1.

References

Gueffrey, D. E. (1986). "Buffers: A Guide for the Preparation and Use of Buffers in Biological Systems." Calbiochem handbook available from Calbiochem Corporation, San Diego, California.

Segal, I. H. (1976). "Biochemical Calculations," 2nd Ed. Wiley, New York.

Questions

1. Show your calculations for preparing the following solutions: 200 ml of 20% (w/v) NaOH, 1 liter of 1.0 M Tris [molecular weight (MW) 121.1 g/mol], and 100 ml of 0.2 M EDTA (MW 372.2 g/mol).

2. How much of the above Tris and EDTA solutions is used to prepare 100 ml of TE buffer (10 mM Tris and 1 mM EDTA)?

3. Describe the relationship between buffer working range and pK value.

4. Discuss the term buffer capacity.

2

Use of Micropipettors and Spectrophotometers

Introduction

Two common laboratory instruments utilized by molecular biologists for the accurate preparation and analysis of many types of samples are micropipettors and spectrophotometers. This exercise is designed to familiarize students with these devices and some of their applications in scientific research.

Micropipettors, when properly used, successfully allow the delivery of small amounts of solution in a variety of desired volumes (see Appendix 5, Use of Micropipettors). This becomes increasingly important in protocols which require accuracy in aliquoting very small increments quickly.

Spectrophotometers have varied applications in the qualitative analysis of sample purity, DNA and protein quantitation, cell density measurements, and assays involving enzyme-catalyzed reactions. Spectrophotometry is based on the simple premise that various compounds will differentially absorb specific wavelengths of light in the ultraviolet (UV, 200–400 nm), visible (VIS, 400–700 nm), or near infrared (near IR, 700–900 nm) ranges. Photometric assays can involve the direct measurement of the absorbance of a sample at a given wavelength or an indirect measurement of an enzymatic reaction product or related substance that serves as an indicator of absorbance directly proportional to the absorbance of the target compound (i.e., assays for protein concentration). Nevertheless, all spectrophotometers employ the same basic structural components designed to detect variations in absorption wavelengths and concentrations.

The general components common to most spectrophotometer systems include a light source, wavelength selector, fixed or adjustable slit, cuvette, photocell, and analog or digital readout (Figure 2.1). Basically, the light source (specific for either UV or VIS ranges) emits light that is passed through the wavelength selector, usually a prism, defraction grating, or set of screening filters, where a specific wavelength of "monochromatic" light is selectively generated (defined by its maximum emission at this wavelength). The light is then directed toward a thin slit (usually adjustable) to regulate its relative intensity before it passes through a cuvette containing the sample of interest. Cuvettes differ with respect to their absorbance characteristics, and one must be careful to be consistent in making all measurements. Finally, absorbance is detected by a photocell which uses the electrons in the refracted light to generate an electrical current that can be amplified and measured to yield an absorbance value, or optical density (OD), for the given sample. The OD reading is corrected against a blank (control: contains all the reagents in the experimental sample except the test compound) to obtain a true measure of the optical density of the sample.

The data obtained from spectrophotometric analyses often serve

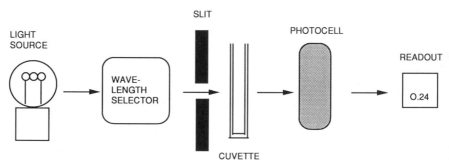

Figure 2.1   Diagram of components of spectrophotometers.

an important role in the determination of various experimental parameters and results. Moreover, when used in conjunction with results from other techniques, spectrophotometric data serve a powerful analytical function.

Reagents/Supplies

Bromphenol blue [1.25% (w/v)]
Cuvettes (alternatively, colorimeter tubes if needed)
Micropipettors and tips
Pipets (10 ml)
Test tubes (18 × 150 mm)

Equipment

Spectrophotometer (if not available, colorimeter can be used)
Vortex

Instructor's Note _____
It will be important to demonstrate to students the proper use of the instrumentation used in your laboratory for this course. This exercise can be done using a variety of instruments to measure optical density. You may wish to have the students compare values obtained from different instruments.

Procedure

Part A. Use of Micropipettors and Spectrophotometers

1. Warm up (about 20 minutes) a Klett colorimeter (green filter, wavelength range 500–570 nm) and/or spectrophotometer set at 540 nm before use. Klett readings can be converted to optical density (OD) by multiplying the Klett reading by 0.002.

2. Place 10 ml of distilled water in each of eight test tubes.

3. With micropipettors add to each successive tube the following amounts of bromphenol blue [1.25% (w/v)]: 0.5, 1, 2, 4, 10, 20, 50, and 100 μl. Manufacturers' instructions for use of micropipettors should be followed scrupulously (see Appendix 5, Use of Micropipettors).

4. Vortex each tube until the dye is in solution.

5. Set the spectrophotometer/colorimeter to zero with distilled water.

6. Transfer the above dye solutions from least concentrated to most concentrated into the same Klett tube or cuvette from reading to reading.

7. Record readings and graph results.

Part B. Determination of Precision of Micropipettors, Operators, and Tips

1. Choose a volume of the bromphenol blue solution (10 μl is suggested) to pipet (in quintuplicate) into each of five test tubes containing 10 ml of distilled water.

2. Read the optical density of each solution as above.

3. Calculate the standard deviation (SD) and coefficient of variation (CV):

$$SD = \frac{\Sigma(X - \overline{X})^2}{N - 1}$$

$$CV = \frac{SD}{X}$$

where X is each individual trial, $\overline{X}$ the arithmetic mean for all five trials, N the number of trials (five), and Σ the symbol for "the sum of."

4. The following comparisons may be made to illustrate variations in the use of micropipettors: (a) the same micropipettor with tips from different manufacturers, (b) different students with the same

micropipettor and tip, and (c) different micropipettors with the same student.

References

Brown, S. B. (1980). "An Introduction to Spectroscopy for Biochemists." Academic Press, New York.

Harris, D. A., and Bashford, C. L. (1987). "Spectrophotometry and Spectrofluorimetry: A Practical Approach." IRL Press, Washington, D.C.

Questions

1. What may account for the difference in OD values obtained using a Klett colorimeter with a green filter as compared to a spectrophotometer set at 540 nm?

2. Consistency of micropipettor usage depends on strict attention to what operational procedures?

3. Explain the relationship among absorbance value, optical density, and percent transmittance.

3

Aseptic Technique: Transferring a Culture

Introduction

Aseptic technique is required to transfer pure cultures and to maintain the sterility of media and solutions (see Appendix 6, Safe Handling of Microorganisms). By aseptic technique the biotechnologist takes prudent precautions to prevent contamination of the culture or solutions by unwanted microbes. Many of the petri dishes and tissue culture plates used for growing pure cultures of microorganisms are made of plastic and come presterilized from the manufacturer; filling these vessels with a sterile medium requires the use of aseptic technique. Proper aseptic transfer technique also protects the biotechnologist from contamination with the culture, which should always be treated as a potential pathogen. Aseptic technique involves avoiding any contact of the pure culture, sterile medium, and sterile surfaces of the growth vessel with contaminating microorganisms. To accomplish this task (1) the work area is cleansed with an antiseptic to reduce the numbers of potential contaminants; (2) the transfer instruments are sterilized (for example, the transfer loop is sterilized by flaming with a Bunsen burner before and after transferring); and (3) the work is accomplished quickly and efficiently to minimize the time of exposure during which contamination of the culture or laboratory worker can occur.

Typical steps for transferring a culture from one vessel to another are (1) flaming the transfer loop, (2) opening and flaming the mouths of the culture tubes, (3) picking up some of the culture growth and transferring it to the fresh medium, (4) flaming the mouths of the culture vessels and resealing them, and (5) reflaming the inoculating loop. Essentially the same technique is used for inoculating petri dishes (except the dish is not flamed) and for transferring microorganisms from a culture vessel to a microscope slide.

Developing a thorough understanding and knowledge of aseptic technique and culture transfer procedures is a prerequisite for working with microbiological cultures. You will save much time and energy and avoid erroneous results if a few simple and commonsense rules are observed when working with cultures.

1. Always sterilize the inoculating loop by flaming before using it to enter any culture material.

2. Always flame the lip of the culture tube before inserting the sterile loop into the culture. This destroys any contaminating cells that may have been inadvertently deposited near the lip of the tube during a previous transfer or by other means.

3. Keep all culture materials covered with their respective caps and lids when not making transfers. Do not lay tube caps or petri dish lids on the tabletop, thereby exposing cultures to possible contamination. When transferring colonies from petri plates, use the lid as a shield by raising it just far enough so that the loop can be inserted but the agar surface is still protected from contaminants falling on it.

4. Do not allow tube closures or petri dish lids to touch anything except their respective culture containers. This will prevent contamination of closures and therefore of cultures.

5. Use proper handling procedures for closure removal and return.

Reagents/Supplies

Inoculating loop

LB broth (Luria–Bertani broth: tryptone, 10 g/liter; yeast extract,
    5 g/liter; and NaCl, 10 g/liter; pH adjusted to 7.5 dropwise with
    NaOH)

Overnight broth cultures of *Escherichia coli* (see Appendix 7, List of
    Cultures)

Test tube racks

Test tubes (18 × 150 mm)

Equipment

Bunsen burner

Incubator at 37°C

Instructor's Note

One day prior to the laboratory exercise, inoculate 5 ml of LB broth
(one culture per two students) with *E. coli*. Grow the culture at 37°C
for approximately 16–24 hours.

Procedure

1. Holding the inoculating loop between your thumb and index
   finger, insert the wire portion into the Bunsen burner flame and
   heat the entire length of the wire until it is red and glowing. Allow
   the wire to cool for 5–10 seconds before immediately proceeding
   to the next step. Do not wave the loop in the air.

2. Using your free hand, pick up the test tube containing the culture
   you want to transfer and gently shake it to disperse the cells.
   Remove the tube cap or plug *with the free little finger of the hand
   holding the sterile inoculating loop* and flame the lip of the tube in
   the Bunsen burner flame.

3. Holding the culture tube in a slanted position, insert the sterile loop and remove a small amount of growth from the tube (a loopful is usually sufficient). Try not to touch the sides of the tube or the sides of the lip with the loop during the removal step.

4. Flame the tube lip again, carefully replace the tube cap, and return the culture tube to the test tube rack.

5. Without flaming the loop, pick up a fresh uninoculated test tube containing 5 ml of LB broth, remove the tube cap or plug, and flame the mouth of the tube.

6. Place the loop containing the viable inoculum into the tube and shake it so that the microorganisms are transferred into the tube.

7. Flame the mouth of the tube, replace the cap or plug, and return the tube to the rack.

8. Flame the inoculating loop.

9. Label the freshly inoculated tube and place it in a 37°C incubator.

10. Repeat Steps 1–9 using sterile LB broth as inoculum in place of the broth culture.

11. At the next laboratory session, examine the tubes and describe their appearance.

Questions

1. Why is it advantageous to hold the culture tube in a slanted position when inserting the loop?

2. What is the function of flaming the tube lip?

4

Establishing a Pure Culture: The Streak Plate

Introduction

The utilization of microbes in biotechnology depends on pure cultures, which consist of only a single species, and the maintenance of the purity of the isolates through subsequent manipulations (see Appendix 8, Storage of Cultures). Most applications in biotechnology involve the use of pure cultures.

Most methods for obtaining pure cultures rely on some form of dilution technique. The most useful and pragmatic approach is the streak plate method in which a mixed culture is spread or streaked over the medium surface such that individual cells become separated from one another. Each isolated cell grows into a colony and, hence, a pure culture because the cells are the progeny of the original single cell.

There are several checks to establish the purity of a culture.

1. On restreaking, an isolated colony from a initial streak plate should yield only a single type of isolated colony whose colonial morphology is consistent with the initial isolate.

2. Microscopic examination of the organisms from a colony should reveal only a single type of cell; differential staining procedures,

such as the gram stain, are useful for establishing that the colony does not contain a mixture of different microbial types.

There are many methods for obtaining a good streak plate, and each method requires some practice. It is important to remember that the more cells you start with on the inoculating loop, the more streaking (dilution) is required. It is not necessary to start with large amounts or gobs of culture material on loops. For this exercise, two different streak plate procedures will be used.

Reagents/Supplies

Inoculating loops

LB agar plates (see Exercise 3 for LB broth preparation; for LB agar, add 15 g/liter of agar to LB broth)

Mixed culture containing *Escherichia coli* and *Saccharomyces cerevisiae*

Pure broth cultures of *Escherichia coli* and *Saccharomyces cerevisiae* (see Appendix 7)

YEPD plates (yeast extract–peptone–dextrose: yeast extract, 10 g/liter; peptone, 20 g/liter; dextrose, 20 g/liter; and agar, 20 g/liter)

Equipment

Bunsen burner
Incubator at 30°C
Incubator at 37°C

Instructor's Note

One day prior to the laboratory exercise, inoculate 5 ml of LB broth with *E. coli* and 5 ml of YEPD broth with *S. cerevisiae* (one culture per student). Grow *E. coli* at 37°C and *S. cerevisiae* at 30°C. On the day of the exercise, prepare mixed cultures by adding 1 ml of *E. coli* and 1 ml of *S. cerevisiae* from the overnight cultures to a sterile test tube.

Procedure

1. Label five LB and five YEPD agar plates with your name, date, class, and the name of the source culture used.

2. Use the *quadrant streak method* (described below) to prepare two streak plates (LB agar and YEPD agar) of the mixed culture. Repeat the procedure using the *continuous streak method* (see below). Also streak by both streaking methods pure cultures of *E. coli* onto LB agar and *S. cerevisiae* onto YEPD agar. Be sure to label each plate.

Quadrant Streak Method (see Figure 4.1)

   a. Draw and number quadrants on the outside bottom of an agar petri plate.

   b. Flame the inoculating loop and be sure to allow the loop to cool before entering the broth culture. Hot loops not only kill micro-organisms but also produce aerosols when they touch the cool agar.

   c. Allow the loop to touch the surface of the agar lightly and slide gently over the surface in one quadrant in a continous streaking motion. Use your petri dish cover to protect the agar surface and to prevent contamination from falling onto the medium. Avoid digging the loop into the agar. Use reflected light to see where you have streaked and inoculated the plate. These areas

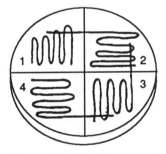

Figure 4.1   Quadrant streak.

will appear as faint scratch marks. This will allow you to better position your inoculations.

d. Flame the loop, allow it to cool, and cross-streak the previous area to produce a second quadrant.

e. Always sterilize the loop after inoculating each section of the plate; this will kill any cells adhering to the loop and prevent contamination of the next inoculation.

f. Repeat the procedure for each succeeding new area, until all four quadrants on the plate are inoculated.

g. Flame the loop when you have finished.

Continuous Streak Method (see Figure 4.2)

a. Take a small amount of culture inoculum on the loop and spread it in a single, continuous, back and forth motion over one-half of the plate.

b. Without flaming the inoculating loop, and using the same face of the loop, turn the plate 180° and continue the streaking procedure as you have done in the initial area.

3. Prepare a control or sham inoculation in which you use only the sterilized inoculating loop without culture to prepare a streak plate by either streaking procedure. The control plate will be a good indicator of your aseptic technique, as nothing should grow on such a plate.

Figure 4.2 Continuous streak.

4. Invert the plates to prevent water condensate from spreading bacteria over the agar surface, place the plates in a 37°C incubator for *E. coli* and LB mixed cultures and in a 30°C incubator for *S. cerevisiae* and YEPD mixed cultures, and incubate for 24–48 hours.

5. At the next laboratory session, observe the streak plates. Observe where you find well-isolated colonies. Compare the colonies you observe in the mixed cultures with standard colony morphologies and with the streak plates of the pure cultures; describe your results.

Questions

1. How can you be certain that you have a pure culture of each organism on each plate?

2. How would you use temperature to enhance or inhibit the growth of bacteria or yeast?

5

Preparation of Culture Media

Introduction

Any medium for the cultivation of bacteria or yeast must provide certain basic nutritional requirements, including (1) a carbon source, which may also serve as an energy source; (2) water; (3) a nitrogen source; (4) a phosphate source; (5) a sulfate source; and (6) various mineral nutrients, such as iron and magnesium. *Escherichia coli* and *Saccharomyces cerevisiae* are capable of growth on a medium consisting of a single carbon source, such as the carbohydrate glucose; a simple nitrogen source, such as ammonium chloride or ammonium sulfate; and other inorganic salts to provide phosphorus, sulfur, and minerals. *Saccharomyces cerevisiae* requires some vitamins as well. This kind of medium is termed defined or synthetic because its exact chemical composition is known. For routine laboratory work, however, complex media in which the basic ingredients are provided by complex nutrients, such as plant and animal extracts of which the exact composition is not known, are often employed. For example, yeast extract and peptone (hydrolyzed protein) are the basic ingredients of LB broth and YEPD. These materials supply a variety of carbon sources, nitrogen compounds in the form of amino acids, and a mixture of cofactors, such as vitamins.

A broth medium is one in which the components are simply dissolved in water. The addition of agar (a complex carbohydrate extracted from red algae) results in a solid medium. Agar is an ideal solidifying agent for microbiological media because of its melting properties and because it has no nutritive value for most bacteria and fungi. Solid agar melts at 90–100°C; liquid agar solidifies at about 42°C.

Sterilization procedures eliminate all viable microorganisms. Culture dishes, test tubes, flasks, pipets, transfer loops, and media must be free of microorganisms before they can be used for establishing pure cultures. The culture vessels must be sealed or capped with sterile plugs to prevent contamination. There are various ways of sterilizing the liquids, containers, and instruments used in pure culture procedures, including exposure to elevated temperatures or radiation levels to kill microorganisms and filtration to remove microorganisms from solution (see Appendix 9, Sterilization Methods). Media preparation usually requires an autoclave for sterilization, which permits exposure to high temperatures for a specific period of time. Generally, a temperature of 121°C (achieved by using steam at 15 psi) for 20 minutes is used to heat sterilize bacteriological media.

In this exercise, you will prepare both a defined and a complex medium. You will learn how to mix the proper constituents to support the growth of *E. coli* and *S. cerevisiae* and how to sterilize the medium.

Reagents/Supplies

Adenine sulfate
Agar
Ammonium chloride (NH_4Cl)
Beakers (1 liter)
Calcium chloride ($CaCl_2$)
Dextrose (glucose) [20% (w/v)]
Erlenmeyer flasks (500 ml and 1 liter)
Graduated cylinder (500 ml)

Heat-resistant gloves
Histidine
Hydrochloric acid (HCl, 1 N)
Lysine
Magnesium sulfate ($MgSO_4 \cdot 7H_2O$)
Pasteur pipets
Peptone
Petri plates, sterile
Potassium phosphate, monobasic (KH_2PO_4)
Screw-cap bottle (100 ml)
Sodium chloride (NaCl)
Sodium hydroxide (NaOH, 1 N)
Sodium phosphate, dibasic (Na_2HPO_4)
Test tube rack
Test tubes (18 × 150 mm) with caps
Tryptophan
Uracil
Yeast extract
Yeast nitrogen base (YNB) with ammonium sulfate, without amino
    acids (Difco)

Equipment

Autoclave
Magnetic stirrer
pH meter
Top-loading balance
Water bath at 50°C

Instructor's Note

Proper budgeting of time is essential for completion of this exercise within the laboratory period. You may wish to divide preparation of various media among groups of students. It is important, however, that the students prepare their own media rather than having this demonstrated to them.

Procedure

Part A. Preparation of YEPD, a Complex Medium for Yeast {Yeast Extract [1% (w/v)], Peptone [2% (w/v)], and Dextrose [2% (w/v)]}

1. Add 500 ml of distilled water to a 1-liter Erlenmeyer flask. Weigh out 5 g of yeast extract, 10 g of peptone, and 10 g of dextrose with a top-loading balance; dissolve these in the water by mixing with a magnetic stirrer. (*Note:* Heating may be necessary.)

2. Divide the broth solution into two equal parts by adding 250 ml each to two 500-ml Erlenmeyer flasks.

   Flask A: Make no further additions.
   Flask B: Add 5 g of agar; swirl to disperse the agar.

3. Dispense the broth media (Flask A) into 15 test tubes, adding 5 ml of broth to each tube.

4. Cap each test tube, but do not tighten excessively, and put aside for autoclaving later. Place Flask B and Flask A (with 175 ml of YEPD remaining) into the autoclave for autoclaving as well.

5. To demonstrate the necessity of the sterilization step, do not autoclave two of the test tubes; simply label and leave them on the benchtop until the next laboratory period. *Before starting Step 6 below, make up M9 medium (see Part B, p. 38) and YNB medium components (see Part C, p. 38).*

6. Load the autoclave with all solutions from Parts A, B, and C requiring autoclaving and close the autoclave door. Most autoclaves have an automatic cycle; set the timer for 20 minutes on the slow exhaust mode. Start the autoclave cycle by pushing the start button in the automatic cycle mode or in the manual mode by turning the selection lever to the fill position. The sterilization cycle involves filling the jacket, allowing steam to enter the chamber, holding the temperature for the amount of time that you have set, and venting the chamber. In the automatic mode, the chamber will

begin to fill with steam after the jacket pressure reaches 15–20 psi; in the manual mode, you must move the selection lever to the fill chamber position when the jacket pressure reaches this level. For manual operation, you must move the selection lever to the vent position after the chamber has been at 121°C for 20 minutes. Slow venting is required to prevent liquids from boiling. Only after complete venting of the chamber can you open the autoclave and remove the material. Autoclaving for 20 minutes actually takes about 40 minutes when you include the time required for heating and venting the chamber. When removing material from the autoclave use heat-resistant gloves: *the material is still hot!*

7. After removal from the autoclave, allow the broth tubes and 175 ml of YEPD (Flask A) to cool before tightening caps and storing for use in later exercises.

8. After the medium in Flask B has been sterilized, place the flask in a 50°C water bath and equilibrate for 30 minutes. (The purpose of the cooling step is to prevent excessive water condensation on the petri dish lid caused by evaporation from agar at temperatures above 50°C as well as to prevent solidification of the agar below 50°C.) To dispense the medium into sterile petri plates, flame the mouth of the flask and, while carefully lifting the lid of a petri plate, pour about 20 ml of agar from Flask B into a plate (enough medium to cover the bottom of the plate). Replace the lid and continue filling additional plates until all of the medium is dispensed. Never completely remove the lids of the petri plates or the plates will become contaminated with bacteria and fungi from the air. Work quickly to minimize contamination, but carefully to prevent accidents. You should periodically reflame the mouth of the flask to reduce contamination. Allow the agar plates to cool. After the agar has solidified, label the plates with your name, date, and medium and leave them at room temperature or incubate at 37°C overnight to allow for drying of the agar and detection of contaminated plates. Store the plates at room temperature in plastic bags or at 4°C for use in later laboratory experiments.

Part B. Preparation of M9 Medium, a Defined Medium for *Escherichia coli* (Glucose–Mineral Salts)

1. Add 450 ml of distilled water to a 1-liter beaker. Weigh out the following amounts of each component and dissolve them in the water:

NH_4Cl	0.5 g
Na_2HPO_4	3.0 g
KH_2PO_4	1.5 g
NaCl	0.25 g

2. Carefully adjust the pH to 7.4 with 1 N NaOH (use a pH meter) and bring the volume to 494 ml using a graduated cylinder. Transfer back to a 1-liter Erlenmeyer flask (Flask C) for autoclaving.

3. Make up 50 ml of the following three solutions in 100-ml screw-cap bottles:

   1 M $MgSO_4 \cdot 7H_2O$
   20% (w/v) dextrose (glucose)
   1 M $CaCl_2$

4. Autoclave all solutions from Steps 2 and 3 above along with Flasks A and B from Part A.

5. At the next laboratory period, aseptically add 1 ml of $MgSO_4 \cdot 7H_2O$ (1 M), 5 ml of glucose [20% (w/v)], and 50 μl of $CaCl_2$ (1 M) to Flask C to make up M9 medium.

Part C. Preparation of YNB Medium, a Defined Medium for *Saccharomyces cerevisiae*

1. Add 6.7 g of yeast nitrogen base (YNB) to 100 ml of distilled water. Slight heating may be required to dissolve. Filter sterilize for a 10× stock solution. Store in a refrigerator.

2. Prepare 100-ml stock solutions of 30 mg/ml lysine, tryptophan, and histidine. Also prepare 100-ml solutions of 2 mg/ml uracil and adenine sulfate. All solutions can be autoclaved except tryptophan

Table 5.1   Preparation of YNB Medium

Component	Medium	
	Liquid	Plates
YNB (10×)	10 ml	10 ml
Agar [4% (w/v)]	—	50
Lysine (30 mg/ml)	0.1	0.1
Tryptophan (30 mg/ml)	0.1	0.1
Histidine (30 mg/ml)	0.1	0.1
Adenine sulfate (2 mg/ml)	1.5	1.5
Uracil (2 mg/ml)	1.5	1.5
Dextrose [20% (w/v)]	10	10
Distilled water	76.7	26.7

and adenine, which should be filter sterilized. (See Appendix 10 for Preparation of Stock Solutions for Culture Media.)

3. Prepare a 100-ml 20% (w/v) dextrose solution, which should be autoclaved.

4. Prepare a 100-ml 4% agar solution. Autoclave. Place molten agar at 50°C to equilibrate for pouring of plates on addition of pre-warmed YNB (see below).

5. Prepare minimal medium with YNB, supplements (amino acids and bases), and dextrose (Table 5.1). For agar plates, add 4% agar as indicated in Table 5.1. Be sure to equilibrate the medium components to 50°C before the addition of agar. Mix well, being careful not to introduce excess bubbles, and pour plates.

References

Sambrook, J., Fritsch, E. F., and Maniatis, T. (1989). "Molecular Cloning: A Laboratory Manual," 2nd Ed. Cold Spring Harbor Laboratory, Cold Spring Harbor, New York.

Sherman, F., Fink, G. R., and Hicks, J. B. (1986). "Laboratory Course Manual for Methods in Yeast Genetics." Cold Spring Harbor Laboratory, Cold Spring Harbor, New York.

Questions

1. Why is the requirement for amino acids much lower than that for glucose in defined media?

2. Explain the usage of both defined and undefined media for the growth of bacteria and yeast.

6

The Growth Curve

Introduction

Knowledge of the growth characteristics of an organism is essential to biotechnology for achieving reproducible transformation efficiencies and for obtaining reproducible plasmid and recombinant protein yields. To obtain uniform, balanced growth, a culture is harvested in the logarithmic or exponential growth phase in which the growth rate and composition of each cell in the population are nearly identical (see Appendix 11, Growth in Liquid Medium, for a full description of the growth curve).

Growth varies in different media, depending on the nutrient level and aeration. Vigorous shaking is necessary to maintain sufficient dissolved oxygen throughout the culture to support growth. Even so, growth in a minimal medium compared to a rich medium will lengthen the doubling time for _Saccharomyces cerevisiae_ and _Escherichia coli_ (Table 6.1). This difference is due to the time and energy the bacteria must spend synthesizing metabolites otherwise supplied in a complex (rich) medium. Thus, the onset and duration of log phase will vary and must be established for the particular medium and strain used. Construction of a growth curve including lag, log, stationary, and death phases will enable you to establish these parameters.

In this exercise, three methods will be used to generate data

Table 6.1   Comparison of Doubling Times in Minimal and Rich Media

Organism	Medium	Doubling time (minutes)
Escherichia coli (strain LE392)	M9 (minimal)	50–60
	LB (rich)	20–30
Saccharomyces cerevisiae (strain YNN281)	YNB (minimal)	200
	YEPD (rich)	90–100

points for constructing a growth curve for *S. cerevisiae* YNN281 in both complex (YEPD) and minimal (YNB) media.

1. Colony counts on agar plates, which measure only *viable* cells (see Appendix 12, Determination of Viable Cells).

2. Changes in the optical density of a culture, which measure *cell mass* (see Appendix 13, Determination of Cell Mass).

3. Direct hemocytometer counts, which measure *cell number* (see Appendix 14, Determination of Cell Number).

Cells from an overnight culture (typically 16–18 hours) are generally in the stationary phase of growth. If the culture is inoculated into fresh medium, the cells will be induced to divide and reenter the log phase (see Appendix 11, Growth in Liquid Medium, Inoculation and Subculture). We will use such a log-phase culture to make measurements.

Reagents/Supplies

Culture of *Saccharomyces cerevisiae* YNN281 (MATa, *ura3-52, trp1-Δ, his3-200, lys2-801, ade2-1*) (see Appendix 7, List of Cultures, and Appendix 15, Nomenclature of Strains)

Cuvettes
Dilution blanks, sterile water in test tubes (16 × 150 mm)
Ethanol [70% (v/v)]
Glass hockey stick spreaders
Hemocytometer and coverslips
Pasteur pipets/bulbs, sterile (9 inches)
Pipets, sterile (1, 5, and 10 ml)
YEPD agar plates
YNB agar plates
YEPD broth (50 ml) in a 250-ml Erlenmeyer flask (for overnight)
YNB broth (50 ml) in a 250-ml Erlenmeyer flask (for overnight)
YEPD broth (250 ml) in a 1-liter Erlenmeyer flask (see Exercise 5)
YNB broth (250 ml) in a 1-liter Erlenmeyer flask (see Exercise 5)

Equipment

Bunsen burner
Incubator at 30°C
Microscope
Rotary shaker at 30°C
Spectrophotometer
Vortex

*Instructor's Note*_____

1. Inoculate *S. cerevisiae* YNN281 into 50 ml of both YEPD and YNB media.
2. Incubate with vigorous shaking overnight at 30°C. The culture should approximate 5×10^7 cells/ml by this time.
3. Dilute into 250 ml of the respective media to $\sim 5 \times 10^3$ cells/ml and incubate at 30°C on a rotary shaker.
4. Remove a 5-ml portion of cells from each culture after 0, 4, 8, 12, 16, 20, 24, 30, and 36 hours of incubation and refrigerate until class time. (The exercise is done this way because continuous monitoring of a culture over 36 hours by the class is not practical.)
5. Draw a chart to be filled in with class data:

Timepoint (hours)	A_{600}		Hemocytometer count		Plate count	
	YNB	YEPD	YNB	YEPD	YNB	YEPD
0						
4						
8						
12						
16						
20						
24						
30						
36						

Procedure

Each pair of students will make measurements on an assigned time-point.

1. Vortex the 5-ml cell suspension of *S. cerevisiae* and aseptically remove a small amount with a sterile Pasteur pipet.

2. Charge the hemocytometer chamber with a drop of culture and count the cells with a microscope (see Appendix 14). The density of the sample may require dilution to make the cells countable by this method. Generally, 200–500 cells in a chamber is convenient.

3. Based on the hemocytometer count, dilute the suspension aseptically using sterile pipets and dilution blanks of sterile distilled water. Yeasts are osmotically stable in water owing to the thick cell wall. Bacteria must be diluted in medium or salt solution. The final tube should contain 500–2000 cells/ml.

4. Plate 0.1 ml of the suspensions in duplicate onto petri plates containing either YEPD or YNB. Spread the cells with a hockey stick sterilized by flaming in ethanol, using a Bunsen burner (see Appendix 12).

5. Incubate inverted plates at 30°C and count colonies in 48 hours.

6. Dispense 1.5 ml of the original cell suspension into cuvettes.

7. Read the A_{600} of the cell suspension in a spectrophotometer against YEPD and YNB blanks for the respective cultures.

Data Collection and Results

1. Fill in the chart with your data.

2. Plot the data of the entire class for insertion into your notebook as follows.

   a. Cell number versus time (plot on linear and semilog paper)

   b. Plate counts (viable cells/ml) versus time (plot on semilog paper)

   c. A_{600} versus time (plot on linear and semilog paper)

   d. A_{600} versus cell number (plot on linear and semilog paper)

References

Ausubel, F. M., Brent, R., Kingston, R. E., Moore, D. D., Seidman, J. G., Smith, J. A., and Struhl, K. (1987). "Current Protocols in Molecular Biology," Chapter 13. Wiley, New York. [Culture of *Saccharomyces cerevisiae*.]

Mandelstam, J., McQuillen, K., and Dawes, I. (1982). "Biochemistry of Bacterial Growth," 3rd Ed., Chapter 2. Halsted, Oxford, England. [Bacterial growth curve.]

Prescott, D. M., ed. (1975). *Methods Cell Biol.* **12**, 1–395.

Questions

1. What are three methods for monitoring cell growth, and what are the parameters measured by each?

2. Calculate the inoculum size (cells/milliliter) for a 5 P.M. inoculation of *S. cerevisiae* YNN281 in YEPD to harvest cells at a density of 3×10^8 cells/ml at 9 A.M. the next morning.

7

Isolation of Plasmid DNA from *Escherichia coli:* The Mini-Prep

Introduction

The mini-prep is a quick method for isolating small amounts of plasmid DNA ($\sim$1 μg/ml of culture) from a transformed host. The mini-prep is much less labor intensive (2 days) than the standard maxi-prep (8 days) which utilizes a cesium chloride density gradient for isolating milligram amounts of plasmid DNA.

The mini-prep is applicable to cloning experiments. Often there are several ways that DNA fragments can come together to form a closed circle which can then transform bacteria. The mini-prep allows rapid screening of transformants by restriction digestion of the small amount of DNA isolated. When the clone of interest is found, large amounts of DNA can then be isolated by the maxi-prep procedure.

Various methods have been developed for rapid, small-scale plasmid isolation. Most notable are the alkaline lysis method (Birnboim and Doly, 1979) and the rapid boiling procedure (Holmes and Quigley, 1981). In this exercise, the latter approach will be used.

Recombinant DNA methodology advances so rapidly that new technologies appear frequently. Several rapid procedures for the mini-prep isolation of plasmid DNA based on chromatographic purification are available commercially from various manufacturers (Qiagen, Pharmacia, 5 Prime → 3 Prime; see Appendix 1, Exercise 9A, Large-Scale Isolation of Plasmid DNA by Column Chromatography). The variety of protocols available enables the molecular biologist to choose among applications suitable to experimental design and budget limitations.

For the purposes of this course, recombinant DNA technology will be introduced in this experiment. In most cases, plasmid DNA will be maintained and carried in a bacterial host. In this exercise, the plasmid pRY121 (see Figure 10.1, Exercise 10) is carried in *Escherichia coli* LE392. Preparation of DNA by the mini-prep method is used to verify the identity of the plasmid DNA. Selective pressure for maintenance of the plasmid is conferred by growth in ampicillin, a penicillin derivative. pRY121 contains a gene encoding β-lactamase, a periplasmic enzyme that cleaves ampicillin thereby yielding cells resistant to this antibiotic. Special care must be taken in the use of glassware for DNA or RNA solutions (see Appendix 16, Glassware and Plasticware).

Reagents/Supplies

Ampicillin stock solution (25 mg/ml in distilled water; filter sterilize and store at $-20°C$)

Ice

Isopropanol (store at $-20°C$)

Lysis buffer [8% (w/v) sucrose, 0.5% (v/v) Triton X-100, 50 mM EDTA (pH 8), and 10 mM Tris-Cl (pH 8) (see Appendix 17, Preparation of Tris and EDTA)]

Lysozyme [10 mg/ml in 10 mM Tris-Cl (pH 8)]

Microcentrifuge tubes, sterile (1.5 ml)

Micropipettors

Overnight culture (5 ml) of *Escherichia coli* LE392 [F⁻ *hsdR514* (r⁻m⁻) *supE44 supF58 lacY1 galK2 galT22 metB1 trpR55* λ⁻]

(pRY121) (see Appendix 7, List of Cultures, and Appendix 15, Nomenclature of Strains)

Pasteur pipets

RNase [DNase-free, 10 mg/ml in 10 mM Tris-Cl (pH 7.5) and 15 mM NaCl; heat to 100°C for 15 minutes, cool slowly to room temperature, store at −20°C]

Sodium acetate (2.5 M)

Styrofoam float

TE buffer [10 mM Tris-Cl and 1 mM EDTA (pH 8)]

Toothpicks

Equipment

Aspirator

Boiling water bath

Microcentrifuge at 4°C

Microcentrifuge at room temperature

Vortex

Water bath at 37°C

Instructor's Note

Grow 5-ml cultures of *E. coli* LE392(pRY121) overnight at 37°C with vigorous aeration in LB medium (see Exercise 3) with ampicillin (50 µg/ml).

Procedure

1. Make up Tris and EDTA stocks as described in Appendix 17. Make TE buffer, lysis buffer, and lysozyme solutions using the Tris and EDTA stocks.

2. Aseptically transfer 1.5 ml of the overnight culture of *E. coli* into a microcentrifuge tube.

3. Centrifuge at top speed (~14,000 rpm) for 1 minute in the microcentrifuge at room temperature.

4. Aspirate and discard spent medium and add another 1.5 ml of culture to the pellet.

5. Centrifuge, discard spent medium, and repeat the procedure with another 1.5 ml of culture. At the last aspiration, leave the pellet as dry as possible but be sure not to aspirate the cells.

6. Save the remaining 0.5 ml of culture and store at 4°C for reference.

7. Resuspend the cell pellet by vortexing in 0.35 ml of lysis buffer. The lysis buffer is designed to weaken the outer membrane of the cell under conditions of osmotic stabilization, providing access of lysozyme to the peptidoglycan layer.

8. Add 25 μl of fresh lysozyme solution and vortex for 3 seconds. The lysozyme solution should be prepared before starting the procedure and should be stored on ice.

9. Place the tube, covered with Parafilm to prevent popping of tube cover, in a styrofoam float in a boiling water bath and boil for 90 seconds. This step serves to inactivate lysozyme.

10. Centrifuge in the microcentrifuge for 10 minutes at top speed at room temperature to pellet the remaining cell envelope with its associated chromosomal DNA.

11. With a toothpick, remove and discard the white, gelatinous pellet from the tube.

12. To the remaining supernatant, add 40 μl of 2.5 M sodium acetate and 420 μl of isopropanol (ice cold) to precipitate DNA. Vortex and store at −20°C for 20 minutes to facilitate precipitation.

13. Centrifuge in the microcentrifuge at top speed for 15 minutes at 4°C.

14. Remove and save the supernatant until DNA precipitation is verified (see Exercise 8). Dry the pellet by inverting the tube on clean absorbent paper and letting it stand for about 20 minutes.

Drying may be hastened by the use of a vacuum desiccator or vacuum centrifuge. Care must be taken with the vacuum desiccator to avoid loss of sample.

15. Resuspend the pellet in 50 μl of TE buffer containing RNase (DNase-free; 50 μg/ml of final concentration in TE). Incubate for 10 minutes in a water bath at 37°C.

16. At this point, the sample may be stored at 4°C until further use.

References

Birnboim, H. C., and Doly, J. (1979). A rapid alkaline extraction procedure for screening recombinant plasmid DNA. *Nucleic Acids Res.* **7**, 1513–1518.

Holmes, D. S., and Quigley, M. (1981). A rapid boiling method for the preparation of bacterial plasmids. *Anal. Biochem.* **114**, 193–197.

Murray, N. E., Brammer, W. J., and Murray, K. (1977). Lambdoid phages that simplify the recovery of *in vitro* recombinants. *Mol. Gen. Genet.* **150**, 53–61. [Description of *E. coli* LE392.]

West, R. W., Yocum, R. R., and Ptashne, M. (1984). *Saccharomyces cerevisiae GAL1–GAL10* divergent promoter region: Location and function of the upstream activating sequence UAS$_G$. *Mol. Cell. Biol.* **4**, 2467–2478. [Construction of pRY121.]

Questions

1. Explain the significance of the following reagents as applicable to the mini-prep procedure: lysozyme, sucrose, RNase, ampicillin, and isopropanol.

2. Show your calculations for preparing the following stock solu-

tions: 20% (w/v) glucose, 0.2 *M* EDTA, 0.75 *M* Tris, and 1 *M* NaOH. How much of the above stock solutions would you use to prepare 100 ml of TE buffer, which is 10 m*M* Tris and 1 m*M* EDTA?

8

Purification, Concentration, and Quantitation of DNA

Introduction

Purification of DNA from a complex mixture of cellular molecules is readily accomplished by removal of proteins and other molecules into an organic solvent. The extraction procedure takes advantage of the properties of phenol and phenol–chloroform which lead to denaturation of proteins. DNA and RNA are not soluble in the organic solvents.

Gel permeation chromatography is a useful procedure for efficient and rapid purification of large molecules, such as proteins and nucleic acids, from small molecules, such as buffer salts and nucleotides. Gels of various porosities are available which exclude molecules of ~10,000–200,000 Da (daltons). For purification of nucleic acids from nucleotides or buffers, an exclusion size of 25,000 Da is convenient. The gels are generally supplied as a dry powder which must be hydrated and poured into columns. Alternatively, hydrated gels may be purchased, or prepoured hydrated gels may be obtained ready for use from a number of manufacturers. A relatively recent innovation in the use of these columns has been centrifugation to speed the separation process. Gel permeation chromatography with centrifugation will be used in Exercise 12 for purification of DNA from nucleotides.

For concentration of purified DNA, the most widely used method is precipitation with ethanol. The precipitated DNA may be recovered by centrifugation and redissolved in a small amount of buffer. In this manner, DNA solutions of a desired concentration can be obtained from very small amounts of DNA, even nanogram levels. Precipitation with ethanol also removes traces of phenol and chloroform which inhibit restriction enzymes and other enzymes used in molecular cloning.

Quantitation of nucleic acids can be carried out by several methods. Nucleic acids absorb UV light at 260 and 280 nm and bind the fluorescent dye ethidium bromide (EtBr). These physical characteristics form the bases of the most convenient and common methods to measure DNA. Furthermore, UV light absorption can be used to assess the purity of the DNA. Spectrophotometric determination is the method of choice if sufficient quantities of relatively pure DNA are to be assayed. Ethidium bromide binding is useful when only small amounts of DNA, or contaminating UV-absorbing material, are present.

PART A   PURIFICATION OF PLASMID DNA BY SOLVENT EXTRACTION

Reagents/Supplies

Chloroform {mixture of chloroform and isoamyl alcohol [24 : 1 (v/v)]}
Ethanol (100%) at −20°C
Microcentrifuge tubes (1.5 ml)
Micropipettors and tips
Phenol (see Instructor's Note, p. 55)
TE buffer (pH 8)

Equipment

Microcentrifuge at room temperature

Instructor's Note

It is important that phenol be free of contaminants which cause DNA damage. The instructor will provide the class with purified phenol prepared according to Sambrook *et al.* (1989). Purified, redistilled phenol can be readily purchased from many companies, however, circumventing the redistillation steps. We recommend using commercially available phenol with the simple additional equilibration with TE buffer according to Sambrook *et al.* (1989).

Procedure

Safety Note

When using phenol, wear gloves and work under a fume hood.

1. Mix 50 μl of the mini-prep DNA sample (Exercise 7) in a microcentrifuge tube with 50 μl of TE buffer (pH 8) to obtain a workable volume.

2. To this mixture add 100 μl of phenol. This organic solvent serves to denature and extract protein.

3. Mix the contents by inverting the tube gently several times until an emulsion forms. This avoids the breakage of DNA that occurs by shear forces generated in vortexing and violent stirring.

4. Centrifuge in the microcentrifuge for 20 seconds at top speed (~14,000 rpm).

5. With a micropipettor, transfer the upper aqueous phase to a clean microcentrifuge tube.

6. Reextract the remaining lower organic phase and interphase in the original tube by adding 100 μl of TE buffer. Mix by inversion. Centrifuge as in Step 4. Collect the aqueous phase and combine with the first aqueous phase collected.

7. Extract the combined aqueous phases by adding about 100 μl of phenol and 100 μl of chloroform. The chloroform also serves to extract and denature protein. Mix by inversion, centrifuge as in Step 4, and transfer the upper phase to a clean microcentrifuge tube.

8. Extract the upper aqueous phase from Step 7 with 200 μl of chloroform only. Mix, centrifuge, and collect the upper phase that contains purified DNA, which should be stored at 4°C.

PART B   CONCENTRATION OF PLASMID DNA

Reagents/Supplies

Ammonium acetate (10 M)
DNA sample from Part A of this exercise
Ethanol (100%) at −20°C
Ice
TE buffer (pH 8)

Equipment

Microcentrifuge at 4°C

Procedure

1. Adjust the concentration of the DNA solution by adding ammonium acetate to make a final solution of 2 M ammonium acetate. This is done by estimating the volume of the DNA sample, which is in TE buffer (pH 8). This is necessary to allow precipitation of DNA by ethanol.

2. Add 2 volumes of ice-cold 100% ethanol. Mix and put on ice for 30–60 minutes. (If the DNA is smaller than 1 kb or is present at less than 100 ng/ml, the solution should be stored at −70°C for about 4 hours. For DNA less than 0.2 kb, the addition of 10 mM MgCl$_2$ improves recovery.)

3. Centrifuge at 4°C for 10 minutes in the microcentrifuge at top speed (~14,000 rpm).

4. Discard supernatant. Invert the tubes on a layer of absorbent

paper to allow drainage of ethanol. Solvent traces can be removed in a vacuum desiccator or vacuum centrifuge.

5. Dissolve the pellet in 50 μl of TE buffer (pH 8). Rinse the tube walls with buffer to ensure dissolution of DNA. Heating to 37°C for at least 5 minutes may help get DNA in solution.

6. Store the DNA solution at 4°C. This mini-prep DNA will be used in Part C of this exercise for quantitation of DNA and in Exercise 10 for restriction digests. Be sure to *save* at least 20 μl for Exercise 10.

PART C   QUANTITATION OF DNA

Reagents/Supplies

*Safety Note*_____
Ethidium bromide is highly toxic. Wear gloves, goggles, and a laboratory coat when working with EtBr (see Instructor's Note, p. 58).

Agarose [1% (w/v), containing 0.5 μg/ml of EtBr]
DNA standards [0.5–20 μg/ml in TE buffer (pH 8); phage λ DNA]
Ethidium bromide stock solution (10 mg/ml; see Instructor's Note, p. 58)
Goggles or face shields, UV blocking
Petri plates
Plastic wrap
Polaroid film (Type 667)
TEA buffer (see Exercise 10)

Equipment

Hand-held shortwave UV light
Polaroid MP-4 land camera (Kodak 22A Wratten filter)
Spectrophotometer
Transilluminator

Instructor's Note_____

For preparation of the EtBr stock solution, add 1 g of EtBr to a 100-ml graduated cylinder. Add 1 ml of 95% (v/v) ethanol and use a magnetic stirrer to mix. The EtBr will dissolve in about 5 minutes. Bring the volume up to 100 ml with distilled water for a 10 mg/ml stock solution. Ethidium bromide is soluble in ethanol and sparingly soluble in water. The solution is light sensitive and should be stored in a brown or foil-wrapped bottle. Store at room temperature. Ethidium bromide is highly toxic; wear gloves, goggles, and a laboratory coat.

Note_____

Inactivation of Ethidium Bromide   For every 100 ml of EtBr solution, add 25 ml of freshly prepared hypophosphorous acid solution and 12 ml of 0.5 *M* fresh sodium nitrite. The final pH of the solution should be less than 3. Stir briefly and let stand for 20 hours. After use of the EtBr-inactivating solution, neutralize with sodium bicarbonate and discard. Inactivation by sodium hypochlorite (Clorox) is not effective in eliminating mutagenicity (Lunn and Sandstone, 1987).

Procedure

Determination by Ultraviolet Absorption

A pure solution of double-stranded DNA at 50 μg/ml has an optical density of 1.0 at 260 nm and an OD_{260}/OD_{280} ratio of 1.8. Contamination with protein or phenol will give a OD_{260}/OD_{280} ratio significantly less than 1.8, and contamination with RNA gives a ratio greater than 1.8. For pure RNA, $OD_{260}/OD_{280} = 2.0$. Thus, the OD_{260}/OD_{280} ratio is obtained first, and, if it approaches 1.8, an accurate estimation of DNA concentration can then be assessed by measuring the absorption at 260 nm.

1. Use DNA standards to construct a standard curve of OD_{260} versus DNA concentration.

2. Dilute the mini-prep DNA (from Part B of this exercise) 1 : 100 in

TE buffer (pH 8) and measure OD_{260} and OD_{280} with a spectrophotometer.

3. Calculate the purity and concentration of the purified mini-prep DNA.

Determination by Ethidium Bromide Fluorescence

If the DNA concentration is less than 250 ng/ml, spectrophotometric assay is not possible. Ethidium bromide binds to DNA by intercalation so that the total fluorescence is proportional to the DNA mass.

1. Stretch commercial plastic wrap over a transilluminator or over a ring, such as the rim of a petri dish with the bottom removed.

2. In an orderly array, spot samples with a micropipettor of your DNA (1–5 μl) and 2 μl of a series of DNA standards (0.5–20 μg/ml, prepared by serial dilution) onto the plastic.

3. Add an equal volume of TE buffer (pH 8) containing 2 μg/ml of EtBr to each spot. Mix with a micropipettor.

4. Illuminate the spots with a shortwave UV light and photograph with a Polaroid camera. You should be able to estimate the concentration of your DNA by comparison to the fluorescence intensities of the known standards.

Safety Note
Ultraviolet light is hazardous and exposure should be minimized. Wear goggles or a face shield.

Determination by Agarose Plate Method

1. Prepare petri dishes containing 10 ml of 1% (w/v) agarose (with 0.5 μg/ml of EtBr) in TEA buffer (see Exercise 10).

2. On the agarose, spot 5 μl of DNA from a stock solution of known concentration (20 μg/ml) as well as serial dilutions in TE buffer

(pH 8) of the stock solution (spot 5 μl of stock and 1 : 2, 1 : 4, 1 : 8, 1 : 16, and 1 : 32 dilutions).

3. Spot 1, 2, and 5 μl of your plasmid DNA on the agarose plate.

4. Allow about 30 minutes for spots to soak into the agarose. This step allows diffusion away from the DNA of contaminating fluorescent or quenching material that may be in the preparation. Leaving the plate overnight will allow more interfering material to diffuse into the agarose.

5. Place the petri dish on the transilluminator and photograph. By comparison of fluorescence intensities you should be able to determine the approximate concentration of DNA in your plasmid preparation.

References

Lunn, E. G., and Sandstone, F. (1987). Ethidium bromide: Destruction and decontamination of solutions. *Anal. Biochem.* **162**, 453–458.

Sambrook, J., Fritsch, E. F., and Maniatis, T. (1989). "Molecular Cloning: A Laboratory Manual," 2nd Ed. Cold Spring Harbor Laboratory, Cold Spring Harbor, New York.

Questions

1. Calculate the amount of DNA in an 150-μl sample with an OD_{260} reading of 0.012.

2. What steps would be taken if your DNA preparation had an OD_{260}/OD_{280} ratio of 1.49 and an OD_{260}/OD_{280} ratio of 1.98?

3. In Part B of this exercise, how manyfold was the DNA concentrated?

9

Isolation of Plasmid DNA: The Maxi-Prep

Note

An alternative method for large-scale isolation of plasmid DNA (given in Appendix 1, Exercise 9A, Large-Scale Isolation of Plasmid DNA by Column Chromatography) substitutes column chromatography for the ultracentrifugation step and takes much less time than the method described below.

Introduction

In practice, clones of transformed bacteria or yeast putatively containing a plasmid of interest can be identified by restriction digests of DNA from mini-preps. After identification, large-scale isolation and purification of plasmid DNA are the next steps for recovering milligram amounts of plasmid. For the purposes of this exercise, the restriction digests and agarose gels will be performed after the maxi-prep so that the qualities of both mini-prepped and maxi-prepped DNAs can be compared.

The maxi-prep entails three basic steps: (1) growing bacteria and amplifying the plasmid, (2) harvesting and lysing the bacteria, and (3) purifying the plasmid from the bacterial host. Selective inhibition of *Escherichia coli* host chromosomal DNA replication with chloramphenicol during the log phase will result in amplification of pRY121 (see Figure 10.1, Exercise 10). Like other plasmids used for cloning, pRY121 has a "relaxed" origin of replication. Relaxed means the plasmid does not require cellular factors for replication; therefore, the plasmid can continue replicating even after growth of the host is arrested. Hundreds of copies of plasmid per cell can be obtained by amplification.

In the second step, the *E. coli* cells will be harvested by centrifugation. Then the cells are weakened by exposure to lysozyme in the presence of sucrose for osmotic stabilization. The cells are finally lysed gently by mixing with a mild detergent–alkali solution. Lysis releases the compact, supercoiled molecules of plasmid into solution, whereas most of the larger chromosomal DNA will remain associated with cellular debris. Because it exists as covalently closed circles, the double-stranded plasmid DNA is not denatured by the alkali.

In the final step, differential centrifugation will yield a pellet containing cellular debris and the majority of cellular DNA. The cleared supernatant is highly enriched for plasmid. Pure plasmid is then isolated from the supernatant by cesium chloride density-gradient ultracentrifugation. The intercalating dye ethidium bromide is added in saturating amounts to the gradient. The dye will bind to nucleic acids and decrease their buoyant density. The purification step here again takes advantage of the circular versus linear nature of the plasmid and host DNAs. The linear *E. coli* DNA will bind more dye per unit length than intact, supercoiled plasmid DNA. Therefore, the plasmid DNA will band below other DNA types on ultracentrifugation to equilibrium. The plasmid band is harvested, ethidium bromide is removed, and the DNA is dialyzed to rid the preparation of cesium chloride. The quality of the plasmid DNA will be tested on an agarose minigel in Exercise 10.

Reagents/Supplies

Ampicillin (25 mg/ml; see Exercise 7)
Cesium chloride (CsCl, "Molecular Biology Grade")

Chloramphenicol (34 mg/ml in 100% ethanol)
Dialysis tubing cut into 10-cm lengths
Distilled water, sterile
EDTA (0.5 *M*, pH 8)
Ethanol [70% (v/v)]
Ethidium bromide (10 mg/ml; see Exercise 8)
Gloves
Glycerol
Goggles or face shields, UV blocking
Isoamyl alcohol
LB broth (see Exercise 3)
Lysozyme [0.02% (w/v) in 8% (w/v) Tris–sucrose]
M9 medium (see Exercise 5)
Needles (18 and 21 gauge)
Oak Ridge polycarbonate centrifuge tubes (30 ml)
Pasteur pipets (6 inches)
Plate of *Escherichia coli* LE392(pRY121)
Polypropylene disposable centrifuge tubes (15 and 50 ml)
Quick-Seal polyallomer centrifuge tubes (13.5 ml)
Scissors
Syringes (5 ml)
TE buffer, 0.1× [1 m*M* Tris (pH 8) and 0.1 m*M* EDTA (pH 8)]
Tris–sucrose [8.5 g of sucrose in 100 ml of 25 m*M* Tris base (pH 8) and 2 m*M* MgCl$_2$]
Triton X-100 solution [0.4% (w/v) Triton X-100, 50 m*M* Tris (pH 8), and 62.5 m*M* EDTA]
Tubing clamps
Vials, cryogenic (optional)

Instructor's Note

Dialysis tubing is supplied as a roll of various widths and must be cut to convenient lengths of 15–20 cm. Tubing should be handled with gloves. The preferred width is 20–25 mm, and the molecular cutoff used is 6000–8000 MW. To prepare the dialysis tubing, fill a 1-liter beaker with 2% (w/v) sodium bicarbonate and 1 m*M* EDTA and bring to a boil. Add about 20 lengths of tubing and boil for 10 minutes. The tubing will float and should be kept submerged. Stirring to promote submerging is discouraged because of the risk of punctures. After boiling, rinse the tubing several times with distilled water by repeated decantation. Boil for 10 minutes in 1 mM EDTA. Store cooled tubing in 0.2 mM EDTA at 4°C. Tubing *must* be submerged or it will dry out.

Equipment

Analytical balance
Balance (for balancing centrifuge tubes)
Beckman J2-21 centrifuge or equivalent (JA-10 rotor)
Beckman Quick-Seal tube sealer or Tube Topper
Beckman ultracentrifuge or equivalent (Beckman SW27 rotor and 80-Ti rotor and caps)
Hand-held shortwave UV light
pH meter
Ring stand and clamps
Shaking incubator at 37°C
Spectrophotometer
Top-loading balance
Water bath at 37°C

Instructor's Note

One day prior to the start of the experiment, inoculate *E. coli* LE392 transformed with pRY121 from an agar plate into 50 ml of LB broth. Add 100 μl of a 25 mg/ml stock of ampicillin to maintain selective pressure for antibiotic resistance. Incubate overnight (~17 hours) at 37°C with vigorous shaking for good aeration of the culture.

Procedure

Day 1

1. Place 1.5 ml of M9 medium into a cuvette to use as a reference for spectrophotometry. You may wish to freeze a sample of the *E. coli*

culture for future reference by adding 0.5 ml of sterile glycerol to 1.0 ml of culture and storing in cryogenic vials at $-70°C$ (see Appendix 8).

2. Inoculate 2.5 ml of the overnight culture into a 1-liter flask containing 250 ml of M9 medium and 0.5 ml of 25 mg/ml ampicillin.

3. Incubate at 37°C with vigorous shaking. Monitor the OD_{600} (growth) of the culture over the next several hours with a spectrophotometer. Make an initial measurement of the optical density directly after inoculation.

4. When the OD_{600} reaches 0.4 (late log phase), add 1.2 ml of chloramphenicol solution. Chloramphenicol will arrest bacterial protein synthesis and allow the plasmids to amplify.

5. Incubate at 37°C with vigorous shaking for 15–18 hours (beyond this time the bacteria will die).

Instructor's Note

On day 2 (prior to the laboratory period).
1. Precool a Beckman J2-21 centrifuge or equivalent and rotor to 4°C.
2. Use 500-ml polycarbonate centrifuge bottles or equivalent to harvest bacteria from Step 5 above. The sterility of the culture beyond this point is no longer critical.
3. Centrifuge the cell suspension at ~4000 g in a Beckman J2-21 centrifuge (or equivalent) at 4°C.
4. Decant the supernatant into a waste flask to be autoclaved. Invert the bottles on a paper towel to drain. Be careful the pellets do not slide out. Pellets should be beige. If the pellets are black, the bacteria are dead. Store pellets on ice. A small amount of dead bacteria is expected.

Day 2

Safety Notes

1. Cesium chloride can *burn* your skin. *Wear gloves!*
2. Ethidium bromide is a *carcinogen*. Handle with *care*. Ethidium

bromide is broken down by treatment with the EtBr-inactivating solution described in Exercise 8. A hand-held UV light can be used to scan the work area and yourself for contamination. The EtBr will fluoresce orange under UV light.

3. Remember that UV light is also hazardous and exposure should be minimized. Wear goggles or a face shield.

4. Dispose of EtBr properly (see Exercise 8).

1. Resuspend the cell pellet prepared by the instructor from overnight growth in chloramphenicol in 3.5 ml of Tris–sucrose solution. Transfer the suspension to a 50-ml disposable polypropylene centrifuge tube. Store on ice.

2. Prepare fresh 0.02% (w/v) lysozyme solution in Tris–sucrose solution. Store on ice. Add 0.8 ml of lysozyme solution per 250 ml of bacterial culture.

3. Add 0.8 ml of 0.5 M EDTA (pH 8.0).

4. Add 4.0 ml of Triton X-100 solution. Invert the tube to mix.

5. Incubate on ice for 1 hour, inverting the tube several times every 20 minutes. This allows time for lysis of bacteria.

6. Pour the lysed bacteria into 30-ml Oak Ridge polycarbonate centrifuge tubes or equivalent. The lysate should be highly viscous and slimy. Balance the tubes against each other by adding Tris–sucrose, if necessary.

7. Centrifuge at 25,000 rpm (Beckman SW27 rotor or equivalent) for 30 minutes at 4°C.

8. Pour the supernatant containing plasmid DNA into a clean 50-ml centrifuge tube. The pellet contains bacterial debris and should be discarded.

9. Note the volume of the supernatant. Add 1 g of CsCl per milliliter of supernatant. Add 0.1 ml of EtBr (10 mg/ml stock) per milliliter of supernatant.

10. Completely dissolve the CsCl by inverting the tube. The ensuing

reaction is highly endothermic. To speed up the dissolution of CsCl, the tube may be placed in a 37°C water bath and inverted every 5 minutes. The total time to dissolve is ~15–20 minutes.

11. Transfer the solution to a 13.5-ml Quick-Seal polyallomer centrifuge tube in the following manner. Remove the plunger from a 5-ml syringe, attach an 18-gauge needle, and put it into the Quick-Seal tube. Use a Pasteur pipet to transfer the solution into the syringe barrel; the solution will drip into the tube.

12. Remove the syringe when the tube is filled and balance the tube against a classmate's tube. A blank of CsCl (1 g/ml) can be used as a balance if there is an uneven number of tubes.

13. Use a heat sealer, such as the Beckman Tube Topper, to seal the tube. Your instructor will explain its use. A *small* bubble should be left in the tube after sealing.

14. Load the tubes into a Beckman 80-Ti rotor, or equivalent, grease the caps and cap rotor slots, and centrifuge for 40–46 hours at 40,000 rpm, 20°C.

Day 4

1. Set up a ring stand with two clamps: one to hold the tube and one to hold a UV lamp.

Safety Note_____

Position a 500-ml beaker containing EtBr-inactivating solution (Exercise 8) under the clamps. Have paper towels on hand. (Cover work area, including walls, with Benchkote.) *Wear gloves and laboratory coats!*

2. Gently remove the tubes from the rotor. Do *not* disturb the gradient. Place the tubes in a centrifuge tube rack.

3. Locate the upper chromosomal and lower plasmid bands under UV light. Each band will fluoresce more brightly than the sur-

rounding solution. Protein is dark purple and floats at the top of the tube. RNA is dark red and is pelleted.

4. Clamp the tube onto the ring stand. Be sure the tubes are secured in the clamp and the ring stand is stable! Locate the bands with UV light.

Safety Note_____

Ethidium bromide may splatter as pressure is relieved. Clean needles in EtBr-inactivating solution and dispose of properly.

5. Use a 21-gauge needle to gently and carefully puncture two to three holes in the top of the tube to break the seal and allow air to enter the tube as the plasmid band is drawn off.

6. Insert a 21-gauge needle on a 5-ml syringe (bevel side up) into the tube just underneath the plasmid band (the *lower* band).

7. *Slowly* draw off the plasmid band. If the bands are observed to be drawing close together during extraction, it is better to leave behind a small amount of plasmid DNA than to contaminate the harvest with chromosomal DNA.

8. Slowly remove the needle containing plasmid DNA. Hold a gloved finger over the hole in the tube until you can submerge it in the beaker filled with EtBr-inactivating solution.

9. Transfer the plasmid DNA to a 15-ml disposable centrifuge tube. Record the volume. Clean the work area with EtBr-neutralizing solution.

Safety Note_____

Perform Steps 10, 11, and 12 *in a fume hood* and avoid inhalation or skin contact of isoamyl alcohol.

10. Mix 30 ml of isoamyl alcohol and 20 ml of sterile distilled water in a 50-ml centrifuge tube. Allow the water to settle to the bottom. Use the upper organic layer to extract the EtBr from the DNA.

11. Add a volume of isoamyl alcohol equal to the DNA volume to the tube. Invert the tube several times and let the aqueous phase containing DNA settle to the bottom.

12. Remove the upper, bright pink organic layer with a Pasteur pipet and discard into a waste bottle. Repeat until the aqueous phase is colorless (~7–8 times), and then once more.

13. Prepare 250 ml of 0.1× TE buffer per preparation for the first dialysis.

14. Wearing gloves, rinse a piece of dialysis tubing (prepared by the instructor) in distilled water; double-knot one end. Allow about 2.5 cm of tubing per milliliter of DNA solution. The tubing should be cut with scissors that have been sterilized by flaming in 70% (v/v) ethanol. Clamp the tubing in front of the knot.

15. Transfer the DNA solution into the tubing with a Pasteur pipet. Do not enter the tubing very far with the Pasteur pipet in order to avoid the risk of puncture. Similarly knot and clamp the open end.

16. Put the tubing containing the DNA solution into 250 ml of 0.1× TE buffer. Stir with a magnetic spin bar on a stir plate at 4°C.

Instructor's Note _____
Change the TE buffer periodically over the next few days. Allow a minimum of 48 hours of dialysis.

Day 7

1. Remove the clamp from one end and cut the dialysis tubing just below the knot with ethanol-sterilized scissors. Hold onto the tubing! Invert a 15-ml disposable polypropylene tube over the opened end. Carefully invert the tubing into the tube and dispense the contents. Alternatively, a sterile Pasteur pipet can be used to *carefully* remove contents.

2. Make a 1 : 10 dilution of the DNA in a microcentrifuge tube: 0.1 ml DNA plus 0.9 ml *sterile* distilled water.

3. Read the optical density of the DNA solution against a sterile distilled water blank in quartz cuvettes.

4. Calculate the DNA concentration, purity, and total yield in milligrams (see Exercise 8).

5. Store the DNA at 4°C.

References

Clewell, D. B., and Helinski, D. R. (1971). Properties of a supercoiled deoxyribonucleic acid–protein relaxation complex and strand specificity of the relaxation event. *Biochemistry* **9**, 4428–4440.

Clewell, D. B., and Helinski, D. R. (1972). Nature of Col E1 plasmid replication in *Escherichia coli* in the presence of chloramphenicol. *J. Bacteriol.* **110**, 667–676.

Questions

1. Compare and contrast the procedures for mini-prep and maxi-prep DNA isolation.

2. Why do chromosomal and plasmid DNA separate on centrifugation in CsCl?

10

Restriction Digestion and Agarose Gel Electrophoresis

Note

Refer to Appendix 18 for Basic Rules for Handling Enzymes.

Introduction

Restriction endonucleases are enzymes that recognize specific nucleotide sequences in double-stranded DNA and are a major tool of the biotechnologist. For prokaryotic cells, they function in nature as restriction–modification systems and will cleave foreign DNA that enters the bacterial cell (e.g., bacteriophage) but not host DNA that has been "protected" or modified by methylation. Restriction enzymes are classified into three systems, Type I, II, and III. For most biotechnology applications, Type II systems are required. Type II systems carry the endonuclease and methylase activities on separate proteins and are useful in molecular cloning. Type II endonucleases will reproducibly cleave DNA at a precise point within a recognition sequence. Generally, different enzymes will recognize different sequences that can be four to six nucleotides long. This precision is essential for molecular cloning techniques, such as isolating genes from genomic DNA or inserting foreign DNA into a plasmid. Many enzymes reproducibly make *staggered* cuts with respect to the dyad axis of symmetry of their recognition sequence. Cleavages of the staggered type will yield DNA fragments with 3′ and 5′ overhangs or "sticky ends" that can base pair with DNA fragments generated by the same enzyme and thus form recombinant molecules. Other enzymes will cut the recognition site *at* the axis of symmetry to yield blunt-ended cleavage products. These can be ligated to other blunt-ended fragments irrespective of the restriction enzyme used.

The second technique introduced in this exercise is agarose gel electrophoresis. When agarose is melted and then cooled in an aqueous solution, it forms a gel via hydrogen bonding. The population of DNA fragments generated by restriction enzymes will move through an agarose gel under the influence of an electric field, where negatively charged DNA molecules will be drawn to the anode. The rate of movement is based almost entirely on size, with the largest molecules having the lowest mobilities. The concentration of agarose in the gel determines the pore size, and a DNA fragment of a particular size will migrate at different rates through gels of different concentrations.

There is a linear relationship between the log of mobility and gel concentration over a certain range of fragment sizes, so a gel concentration must be chosen that will separate the molecules in the DNA population effectively. Gels of 0.8% (w/v) agarose are suitable for separating linear DNA molecules 0.5–10 kb in size. The log molecular weights of known marker fragments, such as λ phage cut with the restriction nuclease *Hin*dIII, can be plotted against mobility. The resulting calibration curve can be used to determine the molecular weights of unknown DNA fragments. As a quick measure, the molecular weight of an unknown fragment can be estimated by direct comparison by eye to the position of λ fragments on the gel. The bands are visualized by UV illumination after staining with the fluorescent dye ethidium bromide.

Purity and intactness of the DNA can also be determined on agarose gels. RNA and protein contamination of maxi-prep DNA may occur during extraction of the plasmid band from the CsCl density gradient. Contamination of mini-prep DNA may occur because of the less rigorous means of purifying the DNA. Contamination with RNA will result in a broad smear running at low molecular weight. RNA can be removed by addition of RNase to restriction digests. Contamination with protein will result in partial inhibition of restriction enzyme activity. In the case of plasmid digests, this results in bands that represent DNA molecules which have not been digested to completion and thus have lower mobilities on the gel. Protein contaminants can be removed by phenol–chloroform extraction and subsequent ethanol precipitation. The agarose gel will also show if the digested plasmid yields the bands expected from known restriction sites on the plasmid map. This serves as a measure of intactness with respect to genetic rearrangement.

In this exercise, we shall use restriction enzymes and agarose minigels as rapid diagnostic tools to determine the size, purity, and intactness of both maxi-prep and mini-prep pRY121 DNA. In practice, DNA fragments can be retrieved from a preparative minigel and be used for cloning. Also, the pattern of fragments can be used for ordering restriction sites on a restriction map.

Reagents/Supplies

Agarose
Bovine serum albumin, fraction V, 10× (BSA, 10 mg/ml)
Dithiothreitol, 20× (DTT, 20 m*M*)
Ethidium bromide (10 mg/ml; see Exercise 8)
λ DNA
λ *Hind*III size markers
Laboratory tape (e.g., Time Tape)
Plastic wrap
Polaroid film (Type 57 or 667) and preservative
pRY121 plasmid DNA from maxi-prep
pRY121 plasmid DNA from mini-prep
Restriction enzyme reaction buffers, 10× (see below)
Restriction enzymes: *Hind*III, *Pst*I, *Bam*HI, and *Eco*RI
Stop buffer [200 m*M* EDTA (pH 8.0)]

TEA buffer, 25× [1 *M* Tris, 15 m*M* EDTA, and 125 m*M* sodium acetate (pH 7.8): to 750 ml of distilled water add 121 g of Tris base, 10.25 g of sodium acetate, and 18.6 g of EDTA; adjust the pH to 7.8 with glacial acetic acid, bring volume to 1 liter, and store at 4°C]

Tracking dye, 6× (see Instructor's Note below)

Tupperware containers (or equivalent) for staining

Equipment

Horizontal minigel electrophoresis apparatus [e.g., Bethesda Research Laboratories (BRL) Baby Gel system]

Microcentrifuge

Microwave oven or Bunsen burner

Polaroid film holder (#545)

Polaroid MP-4 land camera (Kodak 22A Wratten filter)

Power supply capable of delivering 100 V (e.g., BRL Model 100)

Transilluminator

Water bath at 37°C

Water bath at 45°C

*Instructor's Note*_____

To prepare 100 ml of 6× stock solution of tracking dye, add 30 g of glycerol to 70 ml of distilled water. Add 250 mg of bromphenol blue. Store at 4°C. This 6× dye solution is added at one-sixth of the final volume of the DNA solution to be electrophoresed.

Procedure

1. The following guidelines are used to prepare plasmid DNA for restriction digestion.

   a. Determine the quantity of DNA desired to be digested (i.e., 1 μg).

   b. Based on the stock concentration (units/microliter) of the restriction endonuclease, calculate the microliters of enzyme required to achieve digestion of the desired quantity of DNA.

Assume that 2 units of enzyme are required to digest 1 μg of plasmid DNA. (*Note:* 6 units/μg is the rule of thumb for digestion of *chromosomal* DNA.) As use of excess enzyme ensures complete digestion, the manufacturer's stock is often used as supplied without dilution before addition to the reaction mixture.

c. Determine the total volume of the digestion reaction to attain an appropriate volume in which the glycerol content in the reaction will be less than or equal to 5% (w/v). Restriction nucleases are usually supplied in 50% (w/v) glycerol so that a 1:10 dilution of enzyme in the final reaction mixture will suffice.

d. Based on the total reaction volume, calculate how much of both 10× buffer and 20× DTT is required to achieve a 1× final concentration of the reagents in the reaction mixture. For some enzymes, BSA is recommended as well for optimal digestion. The buffer used is dependent on the salt concentration at which the enzyme is optimally active. The appropriate buffer (low, medium, or high salt) is supplied by the manufacturer.

e. Calculate the amount of distilled water required to adjust the volume of the reaction to the desired final volume.

See the sample tabulation below for an example of how such a reaction may be prepared and properly recorded in a notebook.

Tube	Concentration of DNA	DNA solution to give 1 μg of DNA	Enzyme and lot (concentration of enzyme)	Enzyme solution to give 2 U/μg of DNA	10× Buffer (salt concentration)	20× DTT	Distilled water	Total volume
1	0.5 μg/μl	2 μl	*Eco*RI, 61121 (10 U/μl)	1 μl of a 1:5 dilution	2 μl (medium)	1 μl	14 μl	20 μl

2. Calculate the amounts of the ingredients to be used in each of the following restriction digests: (a) λ cut with *Hin*dIII, (b) pRY121 (mini-prep) cut with *Bam*HI, (c) pRY121 (maxi-prep) cut with *Hin*dIII, (d) pRY121 (maxi-prep) cut with *Eco*RI, (e) pRY121 (maxi-prep) cut with *Pst*I, and (f) pRY121 (maxi-prep) cut with *Bam*HI.

Instructor's Note

Manufacturers will supply the recommended buffer with the restriction enzyme. Alternatively, these 10× buffers (referred to as low, medium, or high salt) can be prepared according to the manufacturer's specifications and stored in 1.5-ml aliquots at −20°C. A universal buffer called One-phor-All-Buffer Plus for most restriction endonucleases is available from Pharmacia; it is useful for digestion using multiple enzymes of different salt requirements. It is wise to prepare DTT as a separate 20 mM stock, which can be stored at −20°C in ~25- to 50-μl aliquots and the remainder discarded after use. DTT and BSA are ingredients which should or should not be added to the digest according to the instructions of the enzyme supplier.

3. Use a micropipettor to dispense ingredients into a 1.5-ml microcentrifuge tube. Use a *fresh sterile tip* for each ingredient to avoid cross-contamination. The enzyme should be at −20°C until use and should be added last. All other ingredients should be kept on ice. The enzyme storage buffer is very viscous and will stick to the outside of the tip. To ensure accuracy, the very end of the micropipettor tip should just touch the surface when drawing up the enzyme solution.

4. Mix the digest by gently flicking the microcentrifuge tube. Do not vortex.

5. Pellet any droplets by centrifuging the tube for 3 seconds in a microcentrifuge at top speed.

6. Incubate the digests in a water bath at 37°C for 1 hour. *During this hour,* carry out Steps 7–11 as described on the following page.

7. Assemble the minigel electrophoresis apparatus according to the manufacturer's instructions. The BRL Baby Gel is representative of the many gel systems available and comes with a removable trough in which to form the gel. Tape the ends of the trough and position the trough in the apparatus.

8. Melt 0.2 g of agarose in 25 ml of 1× TEA buffer, using a microwave oven or a Bunsen burner, and equilibrate to 45°C in a water bath.

9. Fill the trough with agarose (45°C) and immediately insert the comb. Act quickly to remove any bubbles by touching them with a Pasteur pipet.

10. Allow the gel to solidify at room temperature for 20 minutes. It will be about 75 mm thick. Gently remove the comb.

11. Remove the tape from the trough and fill the reserviors with 1× TEA buffer to just cover the gel (~400 ml).

12. Pellet the condensation in the tubes by centrifuging for 3 seconds in a microcentrifuge.

13. Add 1 μl of 200 mM EDTA (pH 8) stop buffer. Inactivation of restriction enzymes by chelation of metal cofactors prevents anomalous effects arising from overexposure of DNA to active enzyme. For some restriction endonucleases, other inactivating steps are recommended (see manufacturer's suggestions).

14. Add 4 μl of 6× tracking dye solution. This solution is also very viscous, and the micropipettor plunger should be depressed slowly to release the solution into the digest.

15. Centrifuge the tube for 3 seconds in a microcentrifuge to rid the solution of bubbles.

16. Heat the restriction digest samples at 65°C for 2 minutes to dissociate DNA aggregates.

17. Use a micropipettor to load the restriction digest samples into the wells. It is convenient to correlate tube numbers with the well numbers and order them left to right. Also load 0.5 μg of

λ *Hind*III size markers with 1× tracking dye (available from various suppliers) into the remaining well. Work quickly and efficiently to avoid diffusion of samples.

18. Connect the electrical leads [black = negative (−), red = positive (+)] to the power supply so that the gel runs − to +. Negatively charged DNA molecules will migrate toward the + electrode (anode). Some systems are manufactured with the leads linked to a cover that can attach to the gel box in only one way, thus making the hookup foolproof.

19. Run the gel at 50 V for about 1 hour or until the bromphenol blue dye front is ~1.5 cm from the bottom of the gel.

Safety Note

Do not touch the electrophoresis apparatus with the voltage on. Post *Danger High Voltage* signs.

20. Decant the buffer (reusable) and mark the −/+ orientation by cutting off a corner of the gel. Stain the gel in a Tupperware container for 15 minutes in 0.5 µg/ml of ethidium bromide by adding 5 µl of a 10 mg/ml stock solution of EtBr to 100 ml of distilled water. Recall that EtBr is *highly toxic*.

21. Decant the EtBr into EtBr-inactivation solution (see Exercise 8). Add 100 ml of distilled water to the gel to destain. Agitate gently for 5 minutes.

22. Using gloves, transfer the gel to a transilluminator set up beneath a Polaroid camera and visualize bands under UV light. Following the manufacturer's instructions, load the film cassette into the film holder.

23. Focus the camera on the gel. For convenience, position a ruler next to the gel and focus on the small print. It is important to minimize exposure of both yourself and the gel to UV light. In practice, a restriction fragment may be extracted from the gel for use in cloning experiments, and overexposure to UV light could cause undesirable nicks in the DNA.

24. A red filter (Kodak 22A Wratten filter) should be used over the camera lens. An *f*-stop setting of 4.5 and an exposure setting of B (1 second) give a clear picture of the bands in the gel after developing for 20 seconds.

25. Use the photograph to estimate the size of the fragments by comparison to the λ *Hind*III size markers. Compare the fragments to those expected from the pRY121 restriction map (Figure 10.1). The map is divided into units of kilobase pairs. The correlation of experimental and estimated fragment sizes allows evaluation of the quality of the plasmid preparations.

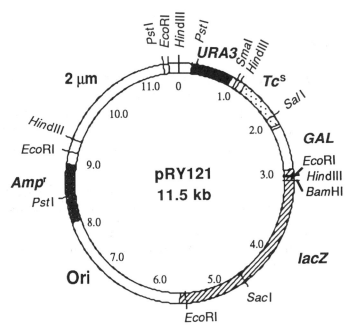

Figure 10.1 Map of plasmid pRY121. Locations of the *Amp*^r (ampicillin resistance), *URA3* (production of uracil), *GAL* (galactose-inducible promoter), and *lacZ* (β-galactosidase) genes are indicated. Ori and 2 μm represent origins of replication for *E. coli* and *S. cerevisiae*, respectively. *Tc*^s corresponds to a truncated gene which no longer confers tetracycline resistance. All cleavage sites for the specified restriction enzymes are also indicated. Consult West *et al.* (1984) for details on the construction of the plasmid.

References

Catalogs of many manufacturers, such as New England Biolabs, Bethesda Research Laboratories, International Biotechnologies, and Stratagene, are excellent sources of information on restriction enzymes and their uses. The manufacturers provide such catalogs on request for distribution to students.

West, R. W., Yocum, R. R., and Ptashne, M. (1984). *Saccharomyces cerevisiae GAL1–GAL10* divergent promoter region: Location and function of the upstream activating sequence UAS$_G$. *Mol. Cell. Biol.* **4**, 2467–2478.

Questions

1. Prepare a reaction mixture for a restriction digest containing components from the following stock solutions: plasmid DNA (3 mg/ml), DTT (20 mM), reaction buffer (10×), and *Bam*HI (10 U/μl). Indicate the minimum total reaction volume required for digesting 15 μg of plasmid DNA and the order in which the reagents are added.

2. Calculate the expected sizes of the DNA fragments from a digest of pRY121 with *Pst*I, with *Hin*dIII, and with *Bam*HI.

3. Explain why cloning into the *Pst*I site at 8.2 kb on pRY121 will lead to a recombinant plasmid that will *not* allow a transformed cell to be ampicillin resistant.

11

Southern Transfer

Introduction

Southern transfer enables the identification of a genetic sequence of interest within a complex mixture of DNA fragments. The DNA sample(s) is digested with restriction enzyme(s) to yield a set of DNA fragments that is then separated according to size by electrophoresis through an agarose gel. The goal of the Southern technique is to obtain a *replica* of the gel which retains the original positions of the DNA fragments. This is accomplished by transferring, or blotting, the DNA from the gel onto a piece of transfer medium. Once immobilized on the transfer medium, the DNA can be probed for the sequence of interest.

In the original protocol of Southern (1975), a piece of nitrocellulose filter paper was used as the transfer medium. More recently, a versatile nylon membrane material has become the preferred medium for many applications. For example, nylon is more amenable to strip-washing, in which the blot is cleared of the original probe by exposure to low salt–high temperature conditions for reprobing. Nylon also does not require baking to fix DNA onto the membrane as does nitrocellulose. This cuts 2 hours off of the procedure time and avoids the problem of brittleness associated with baking. Several brands of membranes are sold by different companies, such as Biotrans (Pall, New York, NY), Nytran (Schleicher and Schuell, Keene, NH), and Zeta Probe (Bio-Rad Laboratories, Richmond, CA).

Blotting is accomplished as follows. Nicks are made in the DNA by exposing the gel to shortwave UV light for 20 seconds. Next the gel is soaked in alkali, which will denature the DNA by breaking hydrogen bonds. The combination of nicking and denaturation is important to facilitate transfer and to provide single strands that may hybridize with complementary single-stranded probe (see Exercise 12). The gel is then neutralized by soaking in Tris buffer, after which it is ready to be blotted. The gel is positioned on top of a buffer-soaked filter, and a piece of transfer medium is placed on top of the gel, completing the gel sandwich. Dry paper towels are then piled on top of the filter. Capillary action pulls the buffer from the bottom filter, through the gel, to the transfer medium, and up through the paper towel stack. A rapid transfer by vacuum has also been developed. The DNA fragments are trapped on the transfer medium, and the DNA is then permanently fixed (to nitrocellulose by baking and to nylon membranes by baking *or* cross-linking the strands with UV light). The replica can then be probed for the sequence of interest.

Reagents/Supplies

Part A. *Agarose Gel Electrophoresis in Vertical Slab Gel Apparatus*

Agarose
Danger High Voltage signs
DNA samples prepared by the instructor (see Instructor's Note, p. 83)
Ethidium bromide (10 mg/ml)
Glass baking dishes (8 × 12 inches, 2 qt)
Micropipettors and tips
Plastic wrap
Polaroid film (Type 57 or 667)

TEA buffer, 25× (see Exercise 10)
Tracking dye, 6× (see Exercise 10)

Instructor's Note_____

1. Gel apparatuses are available from many scientific suppliers at nominal prices. Any apparatus will suffice. If you wish to fabricate your own apparatus, refer to Sambrook *et al.* (1989).
2. Prepare samples of each of the following.
   a. Phage λ DNA cut with *Hin*dIII, 0.5 μg (available from manufacturers)
   b. *Saccharomyces cerevisiae* whole cell DNA cut with *Pst*I, 1 μg (digest 1 μg whole cell DNA from manufacturers with 6 units of *Pst*I)
   c. Human placental whole cell DNA cut with *Pst*I, 1 μg (see Instructor's Note 2b above)
   d. Salmon sperm DNA cut with *Pst*I, 1 μg (see Instructor's Note 2b above)
   e. Plasmid DNA (pRY121) cut with *Pst*I, 1 μg (from student preparation, Exercise 10)

Part B. Southern Transfer

Aluminum foil
Denaturation buffer, 5× (2 *M* NaOH and 4 *M* NaCl)
Ethanol [70% (v/v)]
Glass plates (as weights)
Glass rod or test tube
Neutralization buffer, 1× [0.5 *M* Tris (pH 7.6) and 1.5 *M* NaCl]
Nitrocellulose membranes (0.2 μm) or nylon transfer membranes (0.2 μm)
Paper towels cut to $5\frac{1}{2}$ × $6\frac{1}{2}$ inches (6-inch stack)
Plastic boiling bags for heat sealer
Plexiglass block or glass plates (for platform)
Razor blade
SSC stock solution, 20× [3 *M* NaCl and 0.3 *M* sodium citrate (pH 7)]
Whatman 3 MM filter paper

Equipment

Part A

Microwave oven or Bunsen burner
Polaroid film holder (#545)
Polaroid MP-4 land camera (Kodak 22A Wratten filter)
Power supply capable of delivering 50 V
Transilluminator
Vertical slab gel electrophoresis apparatus
Water bath at 45°C
Water bath at 60°C

Part B

Hand-held shortwave UV light
Heat sealer, such as Sears Seal-A-Meal
Vacuum oven at 70°C (for nitrocellulose)
Water bath at 45°C
Water bath at 60°C

Procedure

Part A. Agarose Gel Electrophoresis in Vertical Slab Gel Apparatus

Follow the guidelines published by the manufacturer of the apparatus used in your class.

Day 1

1. Assemble the gel electrophoresis cell according to the directions of the manufacturer.

2. Prepare 25 ml of 2% (w/v) agarose in TEA buffer (1×) in a 125-ml Erlenmeyer flask. In a 250-ml flask, prepare 100 ml of

1% (w/v) agarose in TEA buffer. Cover with plastic wrap and dissolve in a microwave oven at high power for 3–4 minutes. Watch for boiling and stop immediately. Swirl and hold the flask to the light to inspect for refractile "fish" (undissolved agar particles). Return the flask to the microwave oven and heat until the molten agar is clear by inspection. Alternatively, the agarose can be melted by swirling flasks over a Bunsen burner. Cool to working temperature by placing the 2% agarose in a 60°C water bath and the 1% agarose in a 45°C water bath.

3. Use a 10-ml pipet to dispense quickly a total of 10 ml of 2% agarose down the seam along the inside of each side of the electrophoresis cell. This will seal both sides of the cell and leave a 5-mm plug at the bottom. The agarose will solidify in ~10 minutes. Store the remaining 2% agarose at 60°C for later use.

4. Add the 1% agarose to the cell by tipping back the casting stand and quickly pouring directly from the flask until the level reaches the top of the short glass plate. The agarose should not be above 50°C, or it may shrink during solidification. Below 45°C, the agarose may start to gel and will have to be remelted. Any bubbles that form will usually rise to the top.

5. Immediately insert a gel comb starting at one side and then lowering the other end into the cell. *Avoid bubbles under the comb teeth.* The comb forms the sample wells and will cause some overflow. Allow the gel to solidify at room temperature for 30 minutes.

6. Attach the cell to the electrophoresis apparatus.

7. Create a seal between the cell and the gel apparatus by using a pipet to fill the U-shaped channel at the top with the remaining 2% agarose. Dispense 1–2 ml and repeat several times to fill this space. Allow 5 minutes for the gel to solidify.

8. Fill the upper reservoir with TEA buffer, wait about 2 minutes, and observe for leaks. Fill the lower reservoir. Avoid bubbles between the gel and the lower reservoir. These can be removed using a 21-gauge needle on a 5-ml syringe.

9. Remove the comb slowly and evenly by pulling it upward.

10. Add 6× tracking dye to the restriction digests so that the final concentration of dye is 1×. Heat the DNA samples provided by the instructor for 2 minutes at 65°C.

11. Load 20 μl of each sample using a micropipettor. Work quickly and steadily to minimize sample diffusion which occurs with the voltage off.

12. Hook up the electrical leads to a power supply (black = −, red = +) so the samples run − to +. Run at 25 V for 16 hours.

Safety Note

Do not touch the electrophoresis apparatus with the voltage on. Post *Danger High Voltage* signs.

Day 2

13. The bromphenol blue dye front should be ~1.5–2 cm above the 2% agarose plug at 16 hours. Turn off the power, detach the leads, and decant the buffer. The buffer can be reused several times.

14. Lay the electrophoresis cell flat on nonabrasive Benchkote or other toweling. Gently remove one of the glass plates using a spacer as a lever. Wash the remaining plate and the adherent gel in distilled water. Cut off a small piece of the gel from the upper right corner to mark the orientation.

15. Stain the gel by flooding it for 30 minutes in a 2-qt glass baking dish containing 500 ml of distilled water and 25 μl of a 10 mg/ml stock solution of ethidium bromide. *Wear gloves!*

16. Transfer the gel to another dish of distilled water for 10 minutes to destain. Destaining is facilitated by gentle agitation of the dish.

17. Slide the gel onto a transilluminator. Observe the banding patterns and photograph as in Exercise 10.

Part B. Southern Transfer

Day 2 (Continued)

1. Expose the gel to UV light for 30 seconds to make nicks in the DNA. This allows for better transfer of the DNA to nitrocellulose or nylon membranes.

2. Flood the gel with 400 ml of 1× denaturation buffer in a 2-qt glass baking dish. Shake gently for 30 minutes.

3. Decant the buffer and rinse the gel briefly in distilled water. Flood with 400 ml of 1× neutralization buffer. Shake gently for 20–30 minutes.

4. While waiting during Steps 2 and 3, prepare as follows (see Figure 11.1).

   a. If using nitrocellulose, wet a piece of nitrocellulose corresponding to the size of the gel in distilled water. Handle with gloves. Lower one edge, followed slowly by the rest so it wets evenly. Nylon membranes do not require wetting. Filters may be purchased precut or in sheets or rolls which may be cut to

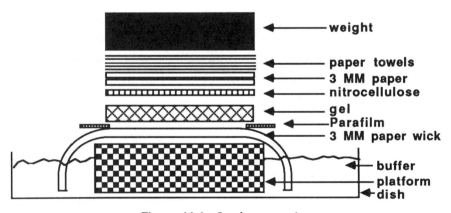

Figure 11.1 Southern transfer.

size with ethanol-sterilized scissors or a paper cutter. Soak filters for at least 5 minutes.

b. Cut pieces of Whatman 3 MM filter paper $\frac{1}{8}$ inch larger than the membrane (five per nitrocellulose, two per nylon membrane). Also cut a wick of Whatman 3 MM paper (see Figure 11.1).

c. Cut four Parafilm strips, 1 × 6 inches, per gel.

d. Cut a 6-inch stack of paper towels to the size of the gel.

e. In a 2-qt glass baking dish, place a piece of plexiglass or a stack of glass plates about 2 inches high. The platform should be about 1 inch longer and wider than the gel. Fill the dish with 10× SSC buffer to a level just below the top of the platform.

f. Wet the wick in 10× SSC and, using a glass rod or test tube, drape it over the platform so both ends are well submerged in buffer. (*Caution:* The wicks tear easily when wet.) Use the glass rod or test tube to roll over the surface to extrude bubbles.

5. Transfer the gel to the top of the wick. Either side can face the membrane. Transfer the nitrocellulose (or nylon membrane) to the top of the gel. Avoid picking up and repositioning the gel because some transfer of DNA occurs immediately.

6. Wet two of the Whatman 3 MM filters together in 10× SSC and position them on top of the membrane. Extrude bubbles from the filters as in Step 4f above.

7. Slip Parafilm strips on the wick along the edges of the gel. The strips will act as barriers to prevent short circuiting of the SSC buffer from the wick to the stack of towels if the towels inadvertently touch the buffer.

8. Stack 3–4 inches of paper towels on top of the filters.

9. Place an evenly distributed 1 $\frac{1}{2}$-pound weight, such as several glass plates, on top of the towels.

10. Cover the whole dish containing the transfer assembly with plastic wrap to prevent evaporation. Allow the transfer to proceed overnight, since fragments larger than 10 kb require ~15 hours to transfer completely.

Day 3

11. Remove the paper towels and Parafilm strips.

12. Pick up the wick–gel–membrane–filter sandwich and invert. Remove the wicks. Mark the orientation of the membrane with a lead pencil by noting the previously cut corner of the gel by well 1. Also mark the well positions. This is important because in practice you may ultimately be measuring the relative mobilities of DNA fragments from an autoradiogram or other detection system.

13. Peel off the gel. Since the gel contains EtBr, it should be treated as a biohazard and discarded accordingly.

14. Since the outermost lanes are not flush with the edges of the gel, several millimeters of transfer medium may be considered excess and can be cut off with an ethanol-sterilized razor blade. This step facilitates fitting the blot into plastic boiling bags (Step 16, p. 90).

15. The DNA must be fixed to the nitrocellulose or nylon filters.

    a. For nitrocellulose, rinse the filters in 400 ml of 2× SSC to remove gel particles and blot the excess fluid onto 3 MM paper. Place the filter between two pieces of Whatman 3 MM paper. Wrap in aluminum foil and bake in a 70°C vacuum oven for 90 minutes. The heating fixes the DNA to the filter. The oven must be under reduced pressure or the nitrocellulose becomes brittle and cracks. After baking, the nitrocellulose is stored until ready to probe.

    b. For nylon membranes, fix the DNA by exposing the side that faced the gel to shortwave UV light for 20 seconds. Rinse in 2× SSC to remove gel particles.

16. The fixed membranes can now be sealed in household plastic boiling bags using a heat sealer to prevent drying until ready to probe. The bags can be purchased presized (e.g., Scotch Kapak) or as a roll (e.g., Seal-A-Meal).

References

Sambrook, J., Fritsch, E. F., and Maniatis, T. (1989). "Molecular Cloning: A Laboratory Manual," 2nd Ed. Cold Spring Harbor Laboratories, Cold Spring Harbor, New York.

Southern, E. (1975). Detection of specific sequences among DNA fragments separated by gel electrophoresis. *J. Mol. Biol.* **98,** 503–512.

Questions

1. What is the purpose of UV-induced nicking and denaturation for facilitating transfer of DNA from agarose gels to nitrocellulose?

2. What are the applications of the Southern transfer procedure to molecular biology?

12

Preparation, Purification, and Hybridization of Probe

Introduction

After DNA has been transferred to nitrocellulose or nylon, detection of the DNA is carried out by hybridization to a probe DNA that is labeled radioactively or derivitized with a detectable group. Nick translation is a process whereby the probe DNA can be modified by incorporation of radioactive or derivatized nucleotides, and it is most commonly used to generate the hybridization probe.

For this exercise, a radioactive probe will not be used because of the regulations and potential hazards associated with radioactive phosphorus, the usual nuclide used for incorporation into probes. If ^{32}P-labeling were used, autoradiography would follow hybridization to detect the transferred DNA–^{32}P-labeled probe hybrid. We will use a nonradioactive alternative which involves incorporation of biotinylated nucleotides into DNA by a standard nick translation protocol, hybridization of the biotin-labeled DNA to the immobilized DNA, and detection of the hybrid by streptavidin and biotin-conjugated alkaline phosphatase (see Figure 12.1). In this detection system, the tetrameric streptavidin binds to the biotinylated probe in such a way that biotin-binding sites remain available to bind the biotin-modified alkaline phosphatase. Addition of a substrate (colorless) for the alkaline phosphatase will indicate the site of

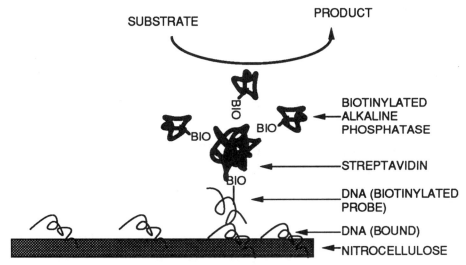

Figure 12.1   Hybridization and detection with biotinylated probe.

the enzyme by production of a product (colored). Thus, a "Dagwood sandwich" is formed *in situ* at the hybrid site that consists of the following: immobilized DNA–biotinylated probe DNA–streptavidin–biotinylated alkaline phosphatase–substrate/product.

In this exercise, we will prepare the biotinylated probe, purify the probe by standard protocols, and hybridize the probe to the Southern blot generated from Exercise 11. After hybridization, the method described above will be used to detect the DNA–DNA hybrid. Many of the reagents used are part of a commercially available kit from Bethesda Research Laboratories (BRL). This system can detect 2–5 pg (picograms) of biotin-labeled probe. This level is sufficient for most probing experiments, and the procedure is sensitive enough to allow detection of a single-copy gene sequence in mammalian DNA. Other nonradioactive alternatives to DNA labeling and detection are available from various manufacturers, such as the Genius system from Boehringer Mannheim and the Visigene system from Promega.

DNA may be conveniently labeled with biotin by nick translation in the presence of biotinylated adenosine triphosphate. The biotin-tagged nucleotide will be efficiently incorporated into DNA by DNA polymerase I in the presence of the other three unlabeled deoxynucleoside triphosphates. The sensitivity of detection of biotin-labeled DNA probes is not greatly affected by the degree of nucleotide incorporation beyond a value of 10–30 biotin nucleotides per kilobase of DNA.

Biotin-labeled DNA may be separated from unincorporated nucleotides by exclusion chromatography (ethanol precipitation also may be satisfactory). Gel filtration on Sephadex G-50 (coarse) is a rapid and quantitative procedure for recovery of labeled probe. Phenol extraction should always be avoided during handling and preparation of biotinylated probes because of possible partitioning of the probe into the phenol layer.

The purpose of passing the nick translation reaction over a gel filtration column is to separate unincorporated nucleotides from the nick-translated probe. Spun columns of Sephadex G-50 offer a fast and easy method to achieve separation of molecules on the basis of size. Sephadex beads contain pores through which molecules under a certain hydrodynamic volume can pass. In this experiment, unincorporated nucleotides are small enough to pass in and out of the beads whereas the probe is not. During the time of centrifugation (hence the name "spun"), the unincorporated nucleotides are held up in the column bead matrix while the probe can pass directly through and be eluted.

The hybridization kinetics of biotinylated DNA probes are virtually identical to those of radioisotope-labeled probes. The rate of solution hybridization is not significantly altered by incorporation of biotinylated nucleotides. However, standard hybridization conditions should be modified for use with biotin-labeled probes owing to the lower melting temperature (T_m) of the probe–target hybrid. Leary *et al.* (1983) modified the technique of Wahl *et al.* (1979) for use with biotin–DNA probes. The steps outlined in Part C of this exercise use these modified protocols.

Instructor's Note_____

An additional application of the biotinylated probe made in this exercise is found in Appendix 1 (see Exercise 12A, Colony Hybridization).

Reagents/Supplies

Part A. Preparation of Probe

Biotin-14-deoxyadenosine triphosphate [0.4 mM in 10 mM Tris (pH 7.5) and 0.1 mM EDTA, BRL 9524SA]

Buffer 1: 0.1 M Tris-HCl (pH 7.5), 0.1 M NaCl, and 2 mM MgCl$_2$

Buffer 2: 3% (w/v) bovine serum albumin (3 g of BSA/100 ml of Buffer 1)

Buffer 3: 0.1 M Tris-HCl (pH 9.5), 0.1 M NaCl, and 50 mM MgCl$_2$

Components of the BRL Nick Translation Reagent System (purchased as a kit) needed for this protocol are as follows

Solution A1: 0.2 mM of dTTP, dGTP, and dCTP in 500 mM Tris-HCl (pH 7.8), 50 mM MgCl$_2$, 100 mM 2-mercaptoethanol, and 100 μg/ml of BSA

Solution B: 5 μg of phage λ DNA in 0.1 mM EDTA, 10 mM Tris-HCl (pH 7.5), and 120 mM NaCl, total volume of 25 μl

Solution C: 0.4 units/μl of BRL DNA polymerase I, 40 pg/μl of DNase I, 50 mM Tris (pH 7.5), 5 mM magnesium acetate, 1 mM 2-mercaptoethanol, 0.1 mM phenylmethylsulfonyl fluoride (PMSF), 50% (w/v) glycerol, and 100 μg/ml of nuclease-free BSA

Solution D (stop buffer): 300 mM disodium EDTA (pH 8.0)

Nick Translation System (BRL 8160SB)

Sodium dodecyl sulfate [5% (w/v) SDS (5 g/100 ml)]

SSC, 20× (see Exercise 11)

Triton X-100 [0.05% (v/v)]

Part B. Purification of Probe

Buffers 1, 2, and 3 (see Reagents/Supplies, Part A)

Corex tubes (15 ml) and cushions

Glass wool, sterile
Microcentrifuge tubes (1.5 ml)
Microcentrifuge tubes (1.5 ml) containing column buffer
Sephadex G-50 in column buffer (see Instructor's Note, p. 97)
Syringes, disposable (1 ml)

Part C. Verification of Biotinylation Reaction

Buffers 1, 2, and 3 (see Reagents/Supplies, Part A)
DNA detection system [BRL 64101, including streptavidin, biotin–
    alkaline phosphatase (AP), nitroblue tetrazolium (NBT), and 5-
    bromo-4-chloro-3-indolyl phosphate (BCIP)]
Pipet tip box tops
Plastic bags
Test strips of nitrocellulose or nylon
Tris [20 mM (pH 7.5)]–EDTA (5 mM) solution

Part D. Hybridization of Probe

Biotinylated probe (from Part B of this exercise)
Denatured salmon sperm DNA (100 μg/ml)
Denhardt's solution, 50× (5 g of Ficoll, 5 g of polyvinylpyrrolidone,
    5 g of BSA fraction V, and distilled water to 500 ml; filter through a
    0.45-μm membrane, dispense in 25-ml portions, and store at
    −20°C)
Glass baking dishes (8 × 12 inches, 2 qt)
Plastic bags for heat sealer
Plastic wrap
Prehybridization buffer, 1.5× (117 g of NaCl, 12.1 g of Tris base,
    0.7 g of disodium EDTA, 2 g of sarkosyl, 2 g of $Na_4P_2O_7 \cdot 10H_2O$,
    and distilled water to 1 liter; adjust the pH to 7.9 with HCl and
    bring to 1.5 liters with distilled water)
Southern blots of various DNA (from Exercise 11)
SSC, 2×, with 0.1% (w/v) SDS
Whatman 3 MM paper

Equipment

Parts A and B

Microcentrifuge
Tabletop centrifuge (Beckman TJ-6 or equivalent)
Water bath at 15°C (this can be set up with a thermostated water bath
in a cold room)

Parts C and D

Hand-held shortwave UV light
Heat sealer
Rotary shaker at 68°C
Transfer plates (disposable incubation trays, such as Accutran from
Schleicher and Schuell)
Water bath at 50°C
Water bath, boiling
Water bath or oven at 68°C

Procedure

Part A. Preparation of Probe

Day 1

1. Using a micropipettor, add the following reagents to a 1.5-ml microcentrifuge tube (which is in ice).

5.0 μl	Solution A1
5.0 μl	Solution B (1 μg λ DNA) or a volume of sample DNA solution giving 1 μg DNA
2.5 μl	0.4 mM biotin-14-deoxyadenosine triphosphate solution
y μl	Water (y = 32.5 μl for λ and is dependent on the volume of sample DNA if sample DNA is used)
45 μl	Total volume

2. Add 5 µl of Solution C. Mix thoroughly but gently by flicking the tube. Centrifuge briefly in a microcentrifuge to bring the liquid to the bottom of the tube.

3. Incubate for 90 minutes in a water bath at 15°C. During this time, complete Steps 1–12 in Part B of this exercise (see below).

4. Add 5 µl of stop buffer (Solution D) and 1.25 µl of 5% (w/v) SDS.

5. Add 50 µl of column buffer (see Instructor's Note). Continue on to Step 13 in Part B of this exercise.

Instructor's Note_____

To prepare a 1.2 g/ml mixture of Sephadex G-50, add 30 g of Sephadex G-50 beads (available from Pharmacia) to 250 ml of sterile column buffer [50 mM Tris (pH 7.9), 2 mM EDTA (pH 8.0), and 250 mM NaCl] in a 500-ml bottle. Allow the beads to swell overnight at room temperature. To hasten swelling, the mixture may be heated to 65°C for 2 hours. Decant the supernatant (allow time for cooling if the latter method is used) and add back an equal volume of column buffer. Store at 4°C. Sephadex G-50 columns, ready for use in DNA purification, are also commercially available.

Part B. Purification of Probe

Day 1 (Continued)

Safety Note_____

Use gloves when handling glass wool.

1. Remove the plunger from a disposable 1-ml syringe.

2. Use the plunger (with rubber tip removed to avoid electrostatic attraction) to tamp a small ball of sterile glass wool to the bottom of the syringe.

3. Replace the rubber tip and retamp to ensure that the wool is adequately compressed.

4. Cut the cap off of a microcentrifuge tube, put the tube into a 15-ml Corex tube, and set it in a rack.

5. Put the 1-ml syringe into the Corex tube so the microcentrifuge tube will catch the column effluent.

6. Add swollen Sephadex G-50 beads to the 1-ml mark on the syringe.

7. Balance the column apparatus.

8. Centrifuge at 1400 g for 4 minutes in a tabletop centrifuge. The Sephadex will pack down.

9. Carefully remove the syringe and discard the contents of the microcentrifuge tube.

10. Reposition the microcentrifuge tube and syringe in the Corex tube.

11. Repeat Steps 6–10 until the level of packed Sephadex reaches the 0.9-ml mark.

12. Run through 100 μl of column buffer. Return to Part A of this exercise, Step 4.

13. Cut the cap off a new microcentrifuge tube and set it up with the column in the Corex tube. Layer the nick translation reaction (in a volume of 100 μl) onto the column and centrifuge as in Step 8 above.

14. Transfer the effluent containing the nick-translated probe into an intact microcentrifuge tube. The purified probe is now ready for use. It can be stored at -20°C.

Part C. Verification of Biotinylation Reaction

Day 2

1. To assure that biotinylated probe is obtained by this procedure, the following steps are taken (use BRL biotinylated DNA for positive control and nonbiotinylated DNA as negative control).

2. Spot 2 μl of a 1 : 5 dilution series of prepared biotinylated pRY121, BRL biotinylated DNA, and BRL nonbiotinylated DNA onto three different test strips (~10 × 0.5 cm) of nitrocellulose paper or nylon. Label each test strip and its orientation with a lead pencil and dry in a vacuum oven at 80°C for 30 minutes. Alternatively, to fix DNA to nylon, expose the nylon strip to shortwave UV light for 20 seconds. *Do not touch filters with fingers.*

3. Wash test strips with Buffer 1 for 1 minute. It is convenient to use transfer plates to ensure the full immersion of the strips in buffer.

4. For nitrocellulose blots, incubate for 20 minutes at 42°C in Buffer 2 (prewarmed to 42°C), and for nylon blots, incubate at 65°C for 45 minutes in Buffer 2 (prewarmed to 65°C).

5. Blot the test strips between two sheets of Whatman 3 MM filter paper and dry overnight at room temperature or in a vacuum oven at 80°C for 15 minutes. The dried strips can be stored desiccated for several months.

Day 3

6. Thoroughly rehydrate the test strips in Buffer 2 for 10 minutes. Drain the buffer.

7. In a polypropylene tube, dilute an appropriate volume of BRL streptavidin to 2 μg/ml by adding 2 μl of a 1 mg/ml stock solution per 1.0 ml of Buffer 1. Prepare ~1.0 ml per test strip. Incubate strips in diluted streptavidin for 10 minutes with gentle agitation, occasionally pipetting the solution over the test strips. Decant the solution.

8. Wash the strips with Buffer 1 using at least a 30-fold greater volume of Buffer 1 than employed in Step 7. A convenient container for this incubation is a pipet tip box top. Gently agitate the strips for 3 minutes in Buffer 1. Decant the solution. Perform this wash step a total of three times.

9. In a polypropylene tube, dilute an appropriate volume of BRL

biotin–AP to 1 μg/ml by adding 1 μl of a 1 mg/ml stock solution per 1.0 ml of Buffer 1. Prepare ~1.0 ml per test strip. Incubate the test strips in diluted biotin–AP for 10 minutes, agitating gently and occasionally pipetting the solution over the strips. Decant the solution.

10. Wash the test strips with Buffer 1 using at least a 30-fold greater volume of Buffer 1 than employed in Step 9. Gently agitate the test strips for 3 minutes in Buffer 1. Decant the solution. Repeat this wash step once.

11. Perform Step 10 using Buffer 3. Repeat this wash step once.

12. In a polypropylene tube, prepare ~7.5 ml of dye solution by adding 33 μl of NBT solution (BRL) to 7.5 ml of Buffer 3, mixing gently (by inverting the tube), and adding 25 μl of BCIP solution, followed by gentle mixing. *The dye solution should be freshly prepared just prior to use.*

13. Incubate the test strips in the dye solution within a sealed plastic bag. Allow the color development to proceed in the dark or in low light for several hours. Maximum color development is usually obtained within 4 hours. Nylon filters may be developed overnight.

14. Wash the test strips in 20 mM Tris (pH 7.5)–5 mM EDTA to terminate the color development reaction. Filters should be stored dry and should always be protected from strong light. To dry, bake at 80°C in a vacuum oven for 1–2 minutes for nitrocellulose or 2–5 minutes for nylon.

Part D. Hybridization of Probe

Day 4

1. Put the dry filter from the Southern blot (Exercise 11) into a plastic bag (e.g., Sears Seal-A-Meal).

2. To each bag, add 15 ml of 1.5× prehybridization buffer, 4 ml of

50× Denhardt's solution, and 0.2 ml of single-stranded, denatured salmon sperm DNA.

3. Squeeze the air out of the bag and seal the end of the bag with a heat sealer. Incubate the bags submerged in a 68°C water bath for 2–4 hours (alternatively, a 68°C oven may be used if available). Occasionally agitate the bags to remove bubbles.

4. Add about 50 μl of purified biotinylated probe to 400 μl of distilled water in microcentrifuge tube. Place the tube in a boiling water bath for 2–4 minutes. Cut open a corner of the bag, add probe, and reseal with heat sealer.

5. Incubate at 68°C overnight.

Day 5

6. Remove the filter from the bag and put it in a 2-qt glass baking dish with 250 ml of 2× SSC with 0.1% (w/v) SDS. Incubate for 3 minutes at room temperature. Repeat this wash. Put the filter in 250 ml of 0.2× SSC with 0.1% SDS for 3 minutes at room temperature. Repeat. Put the filter in 250 ml of 0.16× SSC with 0.1% SDS for 15 minutes at 50°C. Repeat. Briefly rinse the filter in 2× SSC with 0.1% SDS at room temperature.

7. Dry the filter by blotting on Whatman 3 MM paper (do not dry out the filter).

8. Wrap the filter in plastic wrap and store in your desk.

9. To detect hybridized probe, carry out Steps 4 and 7–14 outlined in Part C of this exercise and adjust volumes of detection solutions to 3.0 ml per 100-cm² filter.

References

Anderson, M. L. M., and Young, B. D. (1985). Quantitative filter hybridization. *In* "Nucleic Acid Hybridization: A Practical

Approach" (B. D. Hanes and S. J. Higgins, eds.), pp. 47–111. IRL Press, Oxford, England.

Langer, P. R., Waldrop, A. A., and Ward, D. C. (1981). Enzymatic synthesis of biotin labelled polynucleotides: Novel nucleic acid affinity probes. *Proc. Natl. Acad. Sci. U.S.A.* **78,** 6633–6637.

Leary, J. J., Brigati, D. J., and Ward, D. C. (1983). Rapid and sensitive colorimetric method for visualizing biotin labelled DNA probes hybridized to DNA or RNA immobilized on nitrocellulose: Bioblots. *Proc. Natl. Acad. Sci. U.S.A.* **80,** 4045–4049.

Wahl, G. M., Stern, M., and Stark, G. R. (1979). Efficient transfer of large DNA fragments from agarose gels to diazobenzyloxymethyl paper and rapid hybridization by using dextran sulfate. *Proc. Natl. Acad. Sci. U.S.A.* **76,** 3683–3687.

Questions

1. What are the components of the nick translation reaction? How do they function together to generate a probe?

2. What are the components of the biotinylated DNA detection system, and what is the function of each?

3. How would you change the components of the hybridization reaction to increase or decrease the amount of hybridization?

13

Transformation of *Saccharomyces cerevisiae*

Introduction

Saccharomyces cerevisiae is a valuable organism in the field of bio-technology. It is a eukaryote, yet it can be cultivated like a prokaryote owing to its microbial characteristics. Since *S. cerevisiae* has a small genome and relatively short doubling time and can be analyzed genetically, many of the advances in molecular biology have used this yeast as a research tool. This organism has important industrial uses as well for production of beer, bread, wine, and recombinant peptides and proteins.

The use of *S. cerevisiae* as a host organism for the expression of foreign DNA, introduced in the form of plasmid DNA vectors, has been an important part of current advances in recombinant DNA technology (Botstein and Fink, 1988). Transformation techniques are similar to those applied for *Escherichia coli*, but they are modified to account for the complexity of the yeast cell wall. Transformation of yeast is usually performed in one of two ways: by the formation of spheroplasts, including the removal of the cell wall (Hinnen *et al.*, 1978), or by a more rapid treatment of intact cells with alkali cations (Ito *et al.*, 1983). For our purposes, the latter method will be sufficient for obtaining a desirable number of transformants.

By exposing yeast cells, which have been harvested during the logarithmic phase of growth, to conditions of alkali cations (i.e., LiCl or RbCl), heat shock, and polyalcohol treatment, changes can be induced in the cell envelope that facilitate plasmid uptake. As with *E. coli,* various factors influence the transformation efficiency (see Exercise 14, Introduction). These include plasmid size, DNA configuration, host strain, and selection procedures. Nevertheless, transformation of *S. cerevisiae* is attainable with yields between 10^3 and 10^4 transformants per microgram of DNA for some plasmids.

The pRY121 plasmid used in this course serves as a useful instructional tool. It is a shuttle vector (replicates in two organisms, in this case *E. coli* and *S. cerevisiae*) and encodes an expressible protein, β-galactosidase, which is detectable by a facile assay. Furthermore, the expression of the protein is coupled to a controlling genetic element, the *GAL* promoter. A description of plasmid pRY121 is given in Exercise 10.

Reagents/Supplies

Centrifuge tubes, sterile (50 ml)

LiCl/TE solution (100 mM LiCl in TE buffer)

Microcentrifuge tubes, sterile (1.5 ml)

Polyethylene glycol 4000 [70% (w/v)]

pRY121 DNA, purified by CsCl density-gradient centrifugation or equivalent [from Exercise 9 or 9A (Appendix 1)]

Saccharomyces cerevisiae YNN281

Sorbitol/TE buffer (1 M sorbitol in TE buffer)

TE buffer (pH 8.0; see Exercise 7)

YNB minimal agar plates, selective for transformants (see Part C of Exercise 5, but omit uracil)

YNB minimal agar plates as above with the addition of uracil (30 μg/ml), to use as a control for viability of competent cells

Equipment

Microcentrifuge
Shaking water bath or incubator at 30°C
Tabletop centrifuge or equivalent
Water bath or incubator at 42°C

Instructor's Note

1. Aseptically inoculate 50 ml of YEPD medium in a 250-ml flask with a sterile loop of *S. cerevisiae* YNN281. Incubate with moderate shaking overnight in a water bath or incubator at 30°C. The cell crop from an overnight culture (~12 hours) grown in YEPD should be ~1 × 10^8 cells/ml.
2. The next day, about 4 hours before class, prepare a subculture of the overnight culture by inoculating 100 ml of YEPD medium in a 500-ml flask with an appropriate quantity of cells to achieve a cell density of 1 × 10^7 cells/ml at class time. Assume a doubling time of ~90 minutes in YEPD. Therefore, the density at inoculation should be 1.5 × 10^6 cells/ml. Incubate at 30°C with shaking until class time.

Procedure

1. After the *S. cerevisiae* cells have reached an appropriate density (~1 × 10^7 cells/ml), harvest by centrifuging in 50-ml sterile centrifuge tubes at 2000 rpm for 5 minutes in a tabletop centrifuge.

2. Discard the supernatant. Resuspend the pellet in 50 ml of sterile TE buffer (pH 8.0) and spin at 2000 rpm for 5 minutes.

3. Discard the supernatant. **Gently** resuspend the cell pellet in 30 ml of 100 m*M* LiCl/TE solution. Place in an incubator or water bath for 30 minutes at 30°C with gentle, very slow shaking (*Note:* "Gentle" resuspending should be accomplished by merely

swirling the tube for a few minutes. *Do not bump or invert the tube!)*

4. Centrifuge the suspension at 2000 rpm for 5 minutes.

5. As before, discard the supernatant and gently resuspend the cells in a volume of LiCl/TE solution that gives a density of 5×10^8 cells/ml.

6. Using a micropipettor, transfer 100 μl of competent cells into a sterile 1.5-ml microcentrifuge tube and add 1–10 μg of plasmid DNA to the cells in a maximum volume of 10 μl. Mix gently and incubate without shaking at 30°C for 30 minutes. (Additional 100-μl aliquots of competent cells may be stored at −80°C for reference, if desired.)

7. Add 110 μl of 70% (w/v) polyethylene glycol 4000 solution. Mix gently and incubate at 30°C for 1 hour without shaking.

8. Heat shock the cells by incubating the tube at 42°C for 5 minutes.

9. Immediately pellet the cells by centrifuging in a microcentrifuge for 20 seconds (**not any longer**).

10. Remove the polyethylene glycol-containing supernatant using a micropipettor and resuspend the pellet in 1 ml of sorbitol/TE buffer. (*Note:* It is no longer critical to treat the cells gently; however, it is also not desirable to resuspend them too roughly.)

11. Plate 100, 200, and 400 μl of cells onto minimal agar plates selective for transformants (i.e., YNB agar without amino acids, plus all appropriate supplements **except** uracil for this transformation). In addition, prepare controls by plating untransformed competent cells (from Step 6 above) on the same selective medium and on YNB minimal agar plates with uracil.

12. Allow plates to dry about 20 minutes and incubate, inverted, at 30°C for 3–5 days.

References

Botstein, D., and Fink, G. R. (1988). Yeast: An experimental organism for modern biology. *Science* **240**, 1439–1443.

Hinnen, A., Hicks, J. B., and Fink, G. R. (1978). Transformation of yeast. *Proc. Natl. Acad. Sci. U.S.A.* **75**, 1929–1933.

Ito, H., Fukuda, Y., Kousaku, M., and Kimura, A. (1983). Transformation of intact yeast cells treated with alkali cations. *J. Bacteriol.* **153**, 163–168.

Questions

1. Explain what is meant by transformation efficiency and list five factors that influence it. Describe how each of the factors has its effect(s).

2. Calculate the transformation efficiency achieved in this exercise. Compare this value to that of other students in your class and propose reasons for any differences.

Exercise _____

14

Transformation of *Escherichia coli* by Plasmid DNA

Introduction

The success of many molecular biology protocols is directly related to the ability to achieve efficient uptake of vectors containing the DNA of interest. The most common method of bacterial transformation utilizes high levels of calcium chloride to facilitate the entry of plasmid DNA vectors into *Escherichia coli* by causing structural alterations in the bacterial cell wall. A rapid method using ice-cold 10% (w/v) polyethylene glycol, 5% (w/v) dimethyl sulfoxide (DMSO), and 50 mM magnesium chloride in LB broth has been developed to allow a one-step preparation of competent *E. coli*, which may then be stored until ready for use (Chung *et al.*, 1989). In general, there are three basic steps in the introduction of plasmid DNA into cells: (1) preparation of competent cells, (2) transformation of the competent cells, and (3) selection of transformants.

Most *E. coli* strains can yield a transformation efficiency (measured as number of transformants per microgram of plasmid DNA) between 10^5 and 10^8. The factors influencing this efficiency are often related to conditions that render cells competent, or able to take up

109

DNA. These include the use of cells that are harvested during the logarithmic phase of growth, maintaining cells at a temperature of 4°C during treatment, and prolonged exposure to ice-cold calcium chloride. Following the preparation of competent cells, transformation is induced by destabilization of the lipids in the cell envelope through heat-shock treatment. A subsequent period of recovery is allotted to enable the cells to begin expression of a selectable marker, usually antibiotic resistance. The recovery period is followed by plating of cells onto selective medium containing the selectable marker for which the plasmid-encoded gene product(s) will afford successful transformant survival.

The various parameters and constraints involved in attaining good transformation efficiency differ between cell lines and vectors. Factors, such as plasmid size, DNA configuration (closed, linear, or nicked), and antibiotic/marker selection, may all influence the outcome of a transformation experiment.

Reagents/Supplies

Calcium chloride solution [50 mM CaCl$_2$ and 10 mM Tris-Cl (pH 8.0)]
Centrifuge tubes (10–15 ml)
Escherichia coli LE392
LB agar plates (LB broth plus 20 g of agar/liter)
LB agar plates with ampicillin [LB agar plates plus ampicillin (50 μg/ml); because of its heat instability, ampicillin should be added to the molten agar for plates after cooling to 50°C; the shelf life of the plates at 4°C is 1–2 weeks]
LB broth (see Exercise 3)
LB top agar (LB broth plus 10 g of agar/liter)
LB top agar with ampicillin [LB top agar plus ampicillin (50 μg/ml)]
pRY121 DNA [from Exercise 9 or 9A (Appendix 1)]
TE buffer (see Exercise 7)
Test tubes (13 × 100 mm)

Equipment

Centrifuge at 4°C
Incubator at 37°C
Spectrophotometer
Test tube roller (or shaker) at 37°C
Vortex
Water bath at 42°C

Instructor's Note
1. Aseptically prepare a 5-ml overnight culture of *E. coli* LE392 in LB broth in a test tube on a roller drum (or shaker) at 37°C.
2. The next day (about 3 hours before class) prepare a subculture by transferring 2.0 ml of the overnight cells into a 1-liter flask containing 200 ml of LB broth. Grow the cells with vigorous shaking, to provide proper aeration, in a 37°C water bath to a density of 5×10^7 cells/ml ($OD_{590} = 0.4$). Cell densities above this yield decrease the transformation efficiency.

Procedure

Day 1

1. After the cells have reached the appropriate density, remove 3.0 ml of cells and pipet into a prechilled, sterile centrifuge tube. Chill for 10 minutes in an ice bath.

2. Collect the cells by centrifugation at 4000 rpm for 5 minutes in a centrifuge at 4°C.

3. Carefully discard the supernatant and gently resuspend the pellet in 1.5 ml of sterile, ice-cold 50 m*M* $CaCl_2$ solution. (*Note:* Cultures should be handled gently at all times from this point on.)

4. Centrifuge at 3000 rpm for 5 minutes at 4°C.

5. Discard the supernatant and gently resuspend the pellet in 0.2 ml of ice-cold $CaCl_2$ solution. Store the tubes at 4°C for 12–24 hours.

Day 2

6. Prepare the plasmid DNA for transformation by diluting it to a final concentration of 40 ng/μl in sterile TE buffer.

7. Remove the cells from refrigeration, gently resuspend, and add 10 μl of plasmid DNA. Mix and place on ice for 20 minutes.

8. Heat shock the cells by incubating the tube in a 42°C water bath for 2 minutes.

9. Add 1.0 ml of LB broth, prewarmed to 37°C, to the tube and incubate at 37°C for 1 hour, with occasional gentle mixing, to allow for expression of the selective marker.

10. Dilute the transformed culture (i.e., 1 : 10, 1 : 100, 1 : 1000) using LB broth. Add 150 μl of each dilution to 3.0 ml of LB top agar with ampicillin (equilibrated to 42°C and contained in 13- × 100-mm culture tubes). Working quickly to prevent the top agar from hardening, mix the tube by vortexing and pour onto selective medium (LB agar plates with ampicillin).

11. Plate untransformed, competent cells (Step 5) onto LB agar without ampicillin as a control for viability of competent cells and onto LB agar with ampicillin as a control for transformation efficiency. Be sure to use the correct LB top agar with or without ampicillin in this step.

12. Allow the plates to harden at room temperature for 30 minutes and then incubate at 37°C, inverted, for 12–16 hours. If ampicillin is being used as the selective marker and additional incubation time is desired, plates should be transferred to 4°C after 16 hours to reduce secretion of β-lactamase by transformants that will deplete the antibiotic and result in formation of nonselected colonies, which are often seen as satellite colonies surrounding an ampicillin-resistant colony.

References

Chung, C. T., Niemela, S. L., and Miller, R. H. (1989). One-step preparation of competent *Escherichia coli:* Transformation and storage of bacterial cells in the same solution. *Proc. Natl. Acad. Sci. U.S.A.* **86,** 2172–2175.

Hanahan, D. (1983). Studies on transformation of *E. coli* with plasmids. *J. Mol. Biol.* **166,** 557–580.

Sambrook, J., Fritsch, E. F., and Maniatis, T. (1989). "Molecular Cloning: A Laboratory Manual," 2nd Ed. Cold Spring Harbor Laboratory, Cold Spring Harbor, New York.

Seidman, C. E. (1987). Preparation and transformation of competent cells by using calcium chloride. *In* "Current Protocols in Molecular Biology" (F. M. Ausubel, R. Brent, R. E. Kingston, D. D. Moore, J. D. Seidman, J. A. Smith, and K. Struhl, eds.), pp. 1.8.1–1.8.3. Wiley (Interscience), New York.

Questions

1. Calculate the transformation efficiency obtained in this exercise.

2. Explain how pRY121 can replicate in both *E. coli* and *S. cerevisiae.*

3. Describe the genes and gene products of pRY121 that are used as selectable markers for *E. coli* and *S. cerevisiae.*

Exercise

15

Protein Assays

Introduction

With this exercise, we start a series of experiments designed for familiarization with some basic tools of protein biochemistry. The determination of the concentration of protein in a solution is a critical analytical procedure that must be accomplished rapidly, accurately, with high sensitivity, and at low cost. A variety of procedures have been developed to quantitate protein in biological specimens. Methodologies continue to be improved and new assays developed. Thus, it is difficult to decide which assay(s) to use for a sample. The choice of assay is often based on individual idiosyncrasies of laboratory experience or bias in a particular subfield of biology. In this exercise, we will become acquainted with two assays for protein: the Lowry assay and the Bradford assay. By comparing and contrasting the assays, you may evaluate their relative merits and limitations. The real goal is for you to be prepared to continually adapt proper analytical measurements to your particular application.

A. Lowry Protein Assay

The Lowry reaction for protein determination is an extension of the biuret procedure. The first step involves the formation of a copper–protein complex in alkaline solution. This complex then reduces a phosphomolybdic–phosphotungstate reagent (Folin's reagent) to yield an intense blue color. The Lowry assay is much more sensitive than the biuret method but is also more time consuming. The precaution to be observed when performing the assay concerns addition of Folin's reagent. This reagent is stable only at acidic pH; however, the reduction reaction above occurs only at pH 10. Therefore, when Folin's reagent is added to the alkaline copper–protein solution, mixing must occur immediately so that the reduction can proceed before the phosphomolybdic–phosphotungstate reagent breaks down.

Phenols are capable of reducing molybdenum in a complex of phosphomolybdotungstic acid (Folin–Ciocalteu's reagent). The tyrosine residues of protein provide phenolic groups, and cupric ions enhance the sensitivity. Thus, when treated with Folin–Ciocalteu's reagent, proteins produce color (blue) in varying degrees depending on their tyrosine content. Hence, different proteins give different color values. Generally, the method can be used to determine 20–200 μg of protein per milliliter.

B. Bradford Protein Assay

The Bradford protein assay is a dye-binding assay based on the differential color change of a dye in response to various concentrations of protein. The dye reagent is commonly purchased from Bio-Rad Laboratories. In some research applications, the Bradford assay is recommended as a replacement for other protein assays, especially the widely used Lowry method, for several reasons. First, the Bradford protein assay is much easier to use. It requires one reagent and 5 minutes to perform as compared to three reagents and 30–40 minutes typical for the Lowry assay. Second, because the absorbance of the dye–protein complex is relatively stable, the Bradford assay does not require the critical timing necessary for the Lowry assay. Third, the Bradford assay is not affected by many of the compounds that limit the application of the Lowry assay.

The Bradford protein assay is based on the observation that the absorbance maximum for an acidic solution of Coomassie Brilliant

Blue G-250 shifts from 465 to 595 nm when binding to protein occurs. The extinction coefficient of a dye–albumin complex solution is constant over a 10-fold concentration range. Thus, Beer's law may be applied for accurate quantitation of protein by selecting an appropriate ratio of dye volume to sample concentration. Over a broader range of protein concentrations, the dye-binding method gives an accurate, but not entirely linear, response.

PART A   LOWRY PROTEIN ASSAY

Reagents/Supplies

Bovine serum albumin solutions (0.2, 0.4, 0.6, 0.8, and 1.0 mg/ml)
Folin–Ciocalteu's reagent (dilute 1 : 1 with distilled water so that the final concentration is 1 N)
Reagent A (1.56 g of $CuSO_4 \cdot 5H_2O$/100 ml)
Reagent B (2.37 g of sodium tartrate/100 ml)
Reagent C (2.0 g of NaOH plus 10 g of Na_2CO_3/500 ml)
Reagent D (combine and mix 100 ml of Reagent C, 1 ml of Reagent A, and 1 ml of Reagent B, in this order)
Solution of protein of *unknown* concentration (between 100 and 2000 μg/ml)

Equipment

Spectrophotometer
Vortex

*Instructor's Note*_____

An exercise to demonstrate the effect of various potential contaminants on protein determinations is to add the following to tubes containing 4 ml of a 0.5 mg/ml BSA solution: 80 μl of 10% (w/v) SDS, 40 μl of 20% (w/v) sucrose, 80 μl of 1 M Tris (pH 8), and 16 μl of 0.5 M EDTA (pH 8). A comparison of assay results obtained from these solutions with results from a standard curve of BSA alone is instructive.

Procedure

1. Place 1 ml of Reagent D in appropriately labeled tubes.

2. Add 100 μl of each sample of BSA and various amounts of the unknown solution to the tube of Reagent D and vortex *immediately*. Be sure to include a tube containing no BSA as a blank. Incubate for 10 minutes at room temperature.

3. After the 10-minute incubation, add 100 μl of Folin–Ciocalteu's reagent to the samples and vortex *immediately*. Incubate 30 minutes at room temperature.

4. Measure the OD_{660} of the samples with a spectrophotometer.

5. Plot a curve of OD_{660} (subtract the blank OD_{660} reading from each OD_{660} value for the BSA solutions) versus concentration (milligrams per milliliter) of BSA. Use this standard curve to determine the protein concentration of the unknown.

PART B   BRADFORD PROTEIN ASSAY

Reagents/Supplies

Bovine serum albumin solutions (0.2, 0.4, 0.6, 0.8, and 1.0 mg/ml)
Bradford dye (Bio-Rad 500-0006)
Unknown samples
Whatman filter paper (No. 1)

Equipment

Spectrophotometer
Vortex

Procedure

1. Dilute 1 volume of the Bradford dye with 4 volumes of distilled water.

2. Filter the dye through Whatman No. 1 paper and store at room temperature.

3. Add 2.5 ml of the diluted dye to appropriately labeled tubes.

4. Add 50 μl of BSA and unknown samples to the tubes containing the diluted dye and vortex immediately.

5. Measure the OD_{595} of the samples with a spectrophotometer.

6. Plot the resulting standard curve.

7. Use the standard curve to determine the concentration of the unknown sample.

References

Bradford, M. M. (1976). A rapid and sensitive method for the quantitation of microgram quantities of protein utilizing the principle of protein dye binding. *Anal. Biochem.* **72,** 248–254.

Darbre, A. (1986). Analytical methods. "Practical Protein Chemistry: A Handbook," pp. 227–235. Wiley, New York.

Lowry, O. H., Rosebrough, N. J., Farr, A. L., and Randall, R. J. (1951). Protein measurement with Folin-phenol reagent. *J. Biol. Chem.* **193,** 265–275.

Question

1. Compare and contrast the Lowry and Bradford protein assays with regard to compatibility with various potential contaminating compounds.

16

β-Galactosidase Assay

Introduction

We continue the experiments designed for familiarization with some basic tools of enzyme assay and purification. The series maintains continuity with previous experiments as the enzyme chosen for study is encoded by pRY121. β-Galactosidase is a large enzyme (monomer MW 116,000) that is comparatively stable and easily assayed.

β-Galactosidase (β-D-galactoside galactohydrolase, EC 3.2.1.23) is an enzyme that hydrolyzes β-D-galactosides, such as lactose. It can be measured easily with chromogenic substrates, compounds that yield colored products when hydrolyzed. A commonly used substrate is o-nitrophenyl-β-D-galactopyranoside (ONPG). This colorless compound is converted to galactose and o-nitrophenol in the presence of β-galactosidase. o-Nitrophenol is yellow and can be measured by its absorption at 420 nm. The amount of o-nitrophenol produced is proportional to the amount of enzyme present under conditions of excess ONPG, thereby allowing a quantitative assay for β-galactosidase. For best results, the amount of enzyme should be such that it takes between 15 minutes and 2 hours for a faint color to develop. The reaction is stopped by adding concentrated sodium carbonate solution to shift the pH to 11, at which β-galactosidase is inactive.

Reagents/Supplies

β-Galactosidase (Sigma G7138, crude enzyme from *Aspergillus oryzae*)

o-Nitrophenyl-β-D-galactopyranoside (ONPG, 4 mg/ml)

Sodium carbonate (Na_2CO_3, 1 *M*)

Z buffer (16.1 g of $Na_2HPO_4 \cdot 7H_2O$, 5.5 g of $NaH_2PO_4 \cdot H_2O$, 0.75 g of KCl, 0.246 g of $MgSO_4 \cdot 7H_2O$, and 2.7 ml of 2-mercaptoethanol; adjust to pH 7.0 and bring to 1 liter with distilled water; do not autoclave)

Equipment

Spectrophotometer

Water bath at 28°C

Note_____

See Appendix 18 for Basic Rules for Handling Enzymes.

Procedure

1. Make a 1.0 mg/ml solution of β-galactosidase in Z buffer. Be sure to calculate dilutions so that the units of enzyme in an assay will yield OD_{420} readings between 0.1 and 1.0 within the time frame of your assay (consult the product profile supplied by the manufacturer).

2. Add 100 μl of enzyme solution to 1 ml of Z buffer equilibrated to 28°C. To one tube, add 100 μl of Z buffer alone for a blank of spontaneous hydrolysis of ONPG.

3. Begin the reaction by adding to each tube 0.2 ml of ONPG (4 mg/ml).

4. Incubate at 28°C for 10–30 minutes until a faint yellow color has

developed. If color develops before 10 minutes, dilute the enzyme solution and start over.

5. Stop the reaction by adding 0.1 ml of 1 M Na_2CO_3 and record the length of the incubation time.

6. Read the OD_{420} value against the control with a spectrophotometer.

7. Determine the protein concentration of the enzyme solution (see Exercise 15, Part B).

8. Calculate the specific activity (SA) using the following equation:

$$SA \text{ (units/mg)} = \frac{OD_{420} \times 380}{\text{time (minutes) at } 28°C \times \text{protein (mg)}}$$

The specific activity of $β$-galactosidase is defined in units/mg (1 unit is the amount of enzyme that will hydrolyze 10^{-9} mole of ONPG per minute at 28°C). Pure $β$-galactosidase usually has a specific activity of about 300,000 units/mg. The number 380 in the above equation is the constant used to convert the OD_{420} value to SA units, and it depends on the molar extinction coefficient for *o*-nitrophenol which, under these conditions, is 4500.

Example

Protein concentration in the sample is 10 mg/ml. If 10 μl of a 1 : 100 dilution of the original sample was assayed, there would be 0.01 ml × 0.1 mg/ml = 0.001 mg of protein in the reaction. Given an incubation time of 10 minutes and an OD_{420} reading of 0.4, then

$$SA = \frac{0.4 \times 380}{10.0 \times 0.001} = 15,200 \text{ units/mg}$$

References

Miller, J. H. (1972). "Experiments in Molecular Genetics." Cold Spring Harbor Laboratory, Cold Spring Harbor, New York.

Questions

1. Calculate the molarity of each component in the Z buffer.

2. If 2 mg of BSA were added to 1 ml of your enzyme solution, what would the specific activity of your enzyme solution be?

17

Determination of β-Galactosidase in Permeabilized Yeast Cells

Introduction

Cell permeabilization techniques are often useful for many applications relating to enzyme technology. For example, permeabilization procedures are usually rapid and do not destroy cellular enzymes. Thus, the total amount of an enzyme associated with a cell can be assayed after permeabilization. A number of permeabilization methods for yeast have been developed, such as use of detergents, organic solvents, and desiccation.

In this exercise, *Saccharomyces cerevisiae* cells will be treated with a combination of an organic solvent (toluene) and a detergent (sarkosyl or sodium lauroyl sulfate), which effectively dissolves the permeability barriers (membrane lipids), allowing free access of added substrates to intracellular proteins. Because the cell wall barrier is not broken down by this treatment, however, most proteins remain entrapped and associated with the cell.

Reagents/Supplies

o-Nitrophenyl-β-D-galactopyranoside (ONPG, 4 mg/ml)
Overnight cultures of *Saccharomyces cerevisiae* YNN281 (pRY121)
    and *Saccharomyces cerevisiae* YNN281
Sarkosyl [5% (w/v)]
SF solution [0.85% (w/v) NaCl and 3.7% (w/v) formaldehyde]
Sodium carbonate (Na$_2$CO$_3$, 1 *M*)
Toluene
YNB without amino acids plus supplements as below (Step 1) plus
    galactose (see Exercise 5, Part C)
Z buffer (see Exercise 16)

Equipment

Microcentrifuge
Spectrophotometer
Vortex
Water bath at 28°C
Water bath at 37°C

Procedure

1. Grow *S. cerevisiae* YNN281 at 30°C in 50 ml of liquid 0.67%
   (w/v) YNB medium without amino acids, plus tryptophan
   (30 μg/ml), uracil (30 μg/ml), lysine (30 μg/ml), adenine sulfate
   (30 μg/ml), histidine (30 μg/ml), galactose [3% (w/v)], and glucose [0.2% (w/v)] to mid-log phase (50 Klett units or an OD$_{600}$
   value of 0.7, ~1.7 × 10^7 cells/ml). Grow *S. cerevisiae* YNN281
   (pRY121) the same way but *omit uracil*.

2. Harvest (microcentrifuge for 1 minute) 1.5 ml of each culture in
   microcentrifuge tubes and discard the supernatant. Add another
   1.5 ml of culture to the pellet from the first centrifugation. Repeat
   the centrifugation and resuspend the pellets in 1 ml of Z buffer.

Wash the pellets by centrifugation again and resuspend in 300 μl of Z buffer.

3. Remove 100 μl of resuspended cells and add to 900 μl of SF solution. Determine the OD_{600} with a spectrophotometer.

4. Remove 150 μl of the remaining cell suspension. Add 1.5 μl of toluene and 1.5 μl of sarkosyl. Vortex at top speed several times.

5. Evaporate the toluene for 30 minutes at 37°C on a shaker with the caps off.

6. Add 20 μl of toluenized cells to 1.5 ml of prewarmed (28°C) reaction mixture containing 1 ml of ONPG (4 mg/ml) and 0.5 ml of Z buffer. More cells can be added, with equivalent amounts of Z buffer left out. Vortex and start timing. When the color is yellow or after 120 minutes, add 0.5 ml of Na_2CO_3 (1 M) to stop the reaction.

7. Read the optical density at 420 and 550 nm. The OD_{550} value corrects for light scattering from the cell suspension. Alternatively, the reaction mixture can be centrifuged to pellet cells and the OD_{420} of supernatant read.

8. To calculate units of activity, use the following equation:

$$\text{Units} = \frac{1000 \times (OD_{420} - 1.75\, OD_{550})}{t \times v \times OD_{600}}$$

where t is the assay time (minutes), v is the volume of cells used in the assay (ml), OD_{600} is the cell density at the start of the assay, OD_{420} is the combination of absorbance by *o*-nitrophenol and light scattering by cells, and OD_{550} is the light scattering by cells.

References

Ausubel, F. M., Brent, R., Kingston, R. E., Moore, D. D., Seidman, J. D., Smith, J. A., and Struhl, K., eds. (1988). "Current Protocols in Molecular Biology," pp. 13.6.2–13.6.4. Wiley, New York.

Gowda, L. R., Joshi, M. S., and Bhat, S. G. (1988). *In situ* assay of intracellular enzymes of yeast (*Kluyveromyces fragilis*) by digitonin permeabilization of cell membrane. *Anal. Biochem.* **175,** 531–536.

Questions

1. What are the advantages and disadvantages of enzyme assays on permeabilized cells versus cell extracts?

2. Why is galactose the major carbon source in the growth medium for this exercise?

18

Assay of β-Galactosidase in Cell Extracts

Introduction

A critical step in enzyme purification is extraction of the cellular material from whole cells. The preparation of cell extracts (sometimes called cell-free extracts) must be done with care to ensure the complete release of enzyme from the cellular material without denaturation of the enzyme itself. Often a compromise is made in extraction between complete cell destruction and enzyme recovery because methods used to break open cells are often harmful to proteins.

A variety of methods are used for cell extract preparation: sonication, homogenization with a glass or Teflon pestle, grinding with aluminum or glass beads, and high pressurization and rapid release of pressure (see Table 18.1). For yeast, the most convenient and effective method for cell extract preparation is grinding with glass beads. The shearing action of the glass beads during vortexing ruptures the yeast cells. The very thick and tough cell wall of yeast makes methods, such as sonication, osmotic lysis, and homogenization, ineffective in breaking the yeast cells.

Table 18.1   Cell Extraction Methods

Method	Bacteria	Yeast	Mammalian cells
Sonication	Useful	Not effective	Useful
French press	Useful	Useful only at very high pressures (20,000 psi)	Not useful
Osmotic lysis	Useful for release of periplasmic enzymes	Not effective	Useful
Homogenization with glass–Teflon pestle	Not effective	Not effective	Very useful
Grinding with glass beads	Useful	Effective	Not useful

Reagents/Supplies

Bradford protein assay reagents (see Exercise 15, Part B)
Conical glass tubes (15 ml)
Glass beads (Sigma 2506, 200-μm diameter, acid washed)
Hemocytometer
Log-phase culture of *Saccharomyces cerevisiae* YNN281(pRY121)
o-Nitrophenyl-β-D-galactopyranoside (ONPG, 4 mg/ml)
Phenylmethylsulfonyl fluoride (PMSF, 100 mM; dissolve in 100% isopropanol)
Sodium carbonate (Na_2CO_3, 1 M)
Z buffer (see Exercise 16)

Equipment

Centrifuge
Microscope
Spectrophotometer
Vortex
Water bath at 28°C

Instructor's Note

Grow a 250-ml culture of *S. cerevisiae* YNN281(pRY121) as in Exercise 17 (galactose in YNB plus supplements) for harvest at class time at 1×10^7 cells/ml.

Procedure

1. Harvest a 250-ml culture of *S. cerevisiae* YNN281(pRY121) at late exponential growth phase (1×10^7 cells/ml) by centrifugation at 3000 *g* for 10 minutes. [The number of cells should be determined by OD_{600} measurement or hemocytometer counting prior to harvest (see Appendix 14, Determination of Cell Number). An OD_{600} reading of 0.4 is equal to 1.0×10^7 cells/ml.]

2. Resuspend the cells using conical glass tubes to 2 ml (5×10^8 cells/ml) in Z buffer containing 1 m*M* PMSF. Add glass beads to the height of the meniscus of the cell suspension.

3. Vortex the cells with the glass beads at high speed for 30 seconds and cool on ice for 30 seconds. Repeat vortex and cool for 6 cycles.

4. Let the beads settle, remove and save the cell lysate, add 2 ml of buffer (Z buffer plus 1 m*M* PMSF) to the beads, mix, remove the lysate, and combine lysates. Check the percentage of lysis by counting the cell number with a hemocytometer. Centrifuge the lysate at 5000 *g* for 20 minutes. Collect the supernatant as the cell extract.

5. Determine the protein concentration [using the Bradford protein assay (Exercise 15, Part B)] of the recovered extract and perform the enzyme assay as below.

6. Dilute an aliquot of the lysate at 0.1 mg protein/ml of Z buffer.

7. Add 100 μl of the diluted sample to 0.9 ml of Z buffer and incubate at 28°C for 5 minutes. Run a blank of 1.0 ml of Z buffer only.

8. Begin the reaction by adding 0.2 ml of ONPG (4 mg/ml) which also has been equilibrated in 28°C.

9. Incubate the sample at 28°C until a faint yellow color develops. If color develops immediately after adding the ONPG, dilute the extract or use less extract and start over from Step 7 above.

10. Stop the reaction by adding 0.5 ml of 1 M Na_2CO_3 and record the length of time of incubation.

11. Read the OD_{420}.

12. Calculate the specific activity by using the equation given below (see Exercise 16, Step 8):

$$SA \text{ (units/mg)} = \frac{OD_{420} \times 380}{\text{time (minutes) at 28°C} \times \text{protein (mg)}}$$

References

Catley, B. J. (1988). Isolation and analysis of cell walls. *In* "Yeast: A Practical Approach" (I. Campbell and J. H. Duffus, eds.), pp. 163–184. IRL Press, Washington, D.C.

Pringle, J. R. (1975). Methods for avoiding proteolytic artifacts in studies of enzymes and other proteins from yeasts. *Methods Cell Biol.* **12**, 149–184.

Silverman, S. J. (1987). Current methods for *Saccharomyces cerevisiae*. I. Growth. *Anal. Biochem.* **164**, 271–277.

Questions

1. Compare the specific activities of β-galactosidase obtained from permeabilized cells (Exercise 17) and cell extracts (this exercise). Account for differences that may exist between results from the two protocols.

2. What is the purpose of adding PMSF to the Z buffer (see Pringle, 1975)?

19

β-Galactosidase Purification

Overview

Purification of a protein is essential for a detailed study of its structure and function. In essence, the particular properties of a protein are utilized to separate it from the numerous other proteins in the cell. A purification scheme is usually devised to separate molecular species on the basis of selective solubility, chromatographic behavior, and molecular weight. In cases where these approaches are not sufficient, more sophisticated techniques must also be employed. At every stage of purification, it is important to know the enzyme's "specific activity," that is, the enzymatic activity per milligram of protein, which indicates the degree of purity of the enzyme.

A large-scale culture of *Saccharomyces cerevisiae* YNN281 (pRY121) is harvested, is homogenized by use of glass beads, and is centrifuged to remove the cell debris. Ammonium sulfate is added to this supernatant to precipitate most of the β-galactosidase, which results in a 2- to 5-fold purification. The precipitate is redissolved in buffer, and the ammonium sulfate is dialyzed out against the same buffer. The sample is applied to a gel filtration column followed by elution. The activity peaks are pooled, and the purity of the protein is determined by electrophoresis on sodium dodecyl sulfate–polyacrylamide gels (SDS–PAGE). The sequence used for the purpose of this course is given in Figure 19.1. Other purification steps beyond

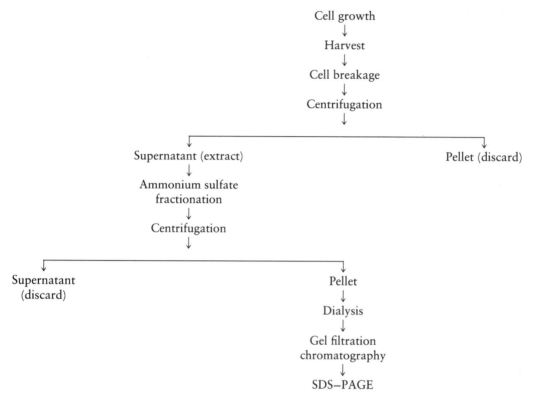

Figure 19.1   Flow chart for the purification of β-galactosidase.

the scope of this introductory course which are often employed include ultrafiltration and chromatography, such as ion-exchange, affinity, and high-performance liquid chromatography.

PART A   GEL FILTRATION CHROMATOGRAPHY: COLUMN CALIBRATION

Introduction

Prior to chromatography of a cell extract, it is necessary to prepare and calibrate the gel filtration column. Thus, we will start the exercise on enzyme purification with column calibration and return to chromatography of the extract following the ammonium sulfate precipitation.

Gel filtration chromatography, also called size exclusion chromatography, impedes the flow of molecules based on their hydrodynamic volume. Separation of molecules is determined by the specific gel filtration medium used and the hydrodynamic volume of the molecules. As hydrodynamic volume is related to molecular size and molecular weight, this chromatography mode can be used to separate proteins. One of the common materials used for gel filtration chromatography is Sephadex, which is a modified dextran cross-linked to give a hydrophilic, three-dimensional network. The degree of swelling is an important characteristic of the gel in which the matrix is a minor component. Loose gels are used for fractionation of high molecular weight substances, whereas more compact gels are used for separation of lower molecular weight compounds.

The choice of an appropriate gel filtration matrix depends on the molecular size and chemical properties of the substances to be separated. Each particular Sephadex type functions mainly within a particular molecular weight range determined by the degree of swelling of the gel. Molecules above a certain molecular weight are totally excluded from particular gels and are eluted in a volume approximately equal to the void volume (volume of interstitial liquid between gel grains in the gel bed). The relationships between elution behavior and molecular properties are given in the applications catalogs of the manufacturers.

Gel filtration yields are usually excellent as there is little or no retention of substances to the column material. A common application of gel filtration is the analysis of mixtures of molecules of different molecular weights. Thus, the determination of molecular weights is a common application of gel filtration chromatography. Other preparative applications of the gel material are desalting (which refers to the removal of salts and other low molecular weight compounds from solutions of macromolecules), concentration of high molecular weight substances, exchange of buffers, removal of organic material (such as phenol from preparations of nucleic acids), removal of low molecular weight, radioactively labeled compounds from solutions of labeled proteins, termination of reactions between macromolecules and low molecular reactants, and removal of products or inhibitors from enzymes. In addition, gel filtration can be used to study chemical equilibria and to separate cells and particles. The student is advised to read gel filtration applications catalogs supplied by the manufacturers.

Reagents/Supplies

Blue Dextran (average MW 2,000,000, Sigma D5751)
Molecular weight standards (Bio-Rad 151-1901 or Sigma MW-GF-200)
Sephadex G-200 (Pharmacia)
Z buffer (see Exercise 16; degas for 5–10 minutes prior to use)

Equipment

Column (2.5 × 50 cm, 250-ml volume)
Fraction collector
Peristaltic pump
UV monitor or spectrophotometer

Instructor's Note
Suspend 6 g of Sephadex G-200 in 500 ml of Z buffer in a 1-liter beaker. Avoid excess stirring that may break beads. Do not use a magnetic stirrer. Place in a boiling water bath for 5 hours. Stir occasionally with a glass rod. Cool to room temperature.

Procedure

Day 1

1. Decant ~60% of the supernatant from above the hydrated Sephadex G-200.

2. Mount the column vertically on a suitable stand. Attach a 500-ml reservoir to the top of the column.

3. Close the column outlet. Add 50 ml of Z buffer. Inject the Z buffer by syringe into the outlet tubing until it passes through the column bed to remove air bubbles.

4. Carefully pour the Sephadex G-200 suspension into the reservoir. The reservoir enables all the suspension to be poured in a single operation.

5. Allow the gel to settle.

6. Open the column outlet and, using a peristaltic pump, adjust the flow rate to not greater than 1 ml/minute.

7. Run at least 2 bed volumes of degassed Z buffer through the gel filtration column in order to allow the bed to stabilize.

8. Poured columns should be protected from contaminating microbial growth during storage. Sodium azide (NaN_3) added to the buffer to 0.02% (w/v) is often used in cases of long-term storage.

Day 2

9. Application and elution of samples is accomplished as follows.

   a. Close the outlet and remove most of the eluent above the gel surface.

   b. Open the outlet to allow drainage of eluent. *Do not allow the bed to run dry.*

   c. To measure the void volume, gently layer Blue Dextran on top of the bed.

   d. Open the column outlet to allow the sample to enter into the gel.

   e. Wash the sample remaining on the bed surface and column wall into the bed with a small amount of Z buffer.

   f. Refill the column with eluent and connect to the pump. In the absence of a pump, the flow rate can be controlled by hydrostatic pressure with the use of a constant pressure flask (see manufacturer's applications manuals).

g. Collect fractions in a suitable volume (i.e., 4-ml fractions every 10 minutes) for assays. The fraction volume must be determined empirically and will differ according to the experimental purification involved. Measure the volume collected until Blue Dextran elutes to determine the void volume.

h. Repeat the elution (Steps c–g above) with molecular weight markers. If colored markers are used, simply determine the fraction number or elution volume (milliliters) where the marker elutes from the column. If noncolored markers are used, read the column eluent by OD_{280} with a UV monitor or read each fraction at OD_{280} with a spectrophotometer. Plotting the molecular weights of the marker proteins versus elution volume yields a calibration curve for the column.

PART B   AMMONIUM SULFATE SALTING-OUT OF β-GALACTOSIDASE AND COLUMN CHROMATOGRAPHY

Introduction

In cell extracts, the desired protein is often less than 1% of the total protein. In this case, differential solubility procedures are generally used as the first step to separate a protein of interest from "contaminating" proteins. Differential solubility procedures have a high capacity to eliminate gross impurities. Methods of higher resolution are used in subsequent purification steps.

Heat treatment, selective precipitation (precipitation of protein at its pI in a solution of pH = pI), organic solvent precipitation, polyethylene glycol precipitation, and salt fractionation are all differential solubility techniques. However, the most useful in the majority of cases is salt fractionation with ammonium sulfate. Ammonium sulfate is highly soluble in water, available at high purity and low cost, and stabilizes proteins.

At low ionic strengths, addition of ammonium sulfate has a "salting-in" effect of enhancing protein solubility. With addition of more ammonium sulfate, protein becomes less soluble and undergoes "salting-out," or precipitation. The concentration of ammonium sulfate at which each protein precipitates is characteristic for that protein and varies with pH, temperature, and protein concentration. Proteins precipitated by ammonium sulfate usually retain their native conformation, can be redissolved readily, and can be stored in precipitated form.

Instructor's Note
It will be necessary for this exercise to prepare a large-scale cell extract of *S. cerevisiae* YNN281(pRY121). Follow the procedures of Exercise 18, using 2–4 liters of cell culture that has been induced to produce β-galactosidase by growth in galactose.

Reagents/Supplies

Ammonium sulfate [$(NH_4)_2SO_4$, as pure as possible]
Cell extract (prepared by the instructor)
Dialysis tubing (see Instructor's Note below)
Materials for β-galactosidase assay (see Exercise 16)
Materials for protein assay (see Exercise 15)
Phenylmethylsulfonyl fluoride (PMSF; see Exercise 18)
Sodium hydroxide (NaOH)
Z buffer (see Exercise 16)

Equipment

Calibrated columns (see Part A of this exercise)
Centrifuge at 4°C
Cold room or ice bath
Magnetic stirrer

Instructor's Note
Prepare dialysis tubing as in Exercise 9. For this exercise, use Visking #20 (Union Carbide). The #20 tubing is the most porous of the

Visking series, thus dialysis times can be relatively short. For small samples (as little as 50 μl can be dialyzed in #20 sacs), very rapid equilibrium is attained owing to the large surface area available.

Procedure

Day 1

1. Dilute the cell extract to 100 ml with Z buffer in a 250-ml beaker. Place on a stir plate in an ice bath. Add the stirrer to the extract solution and rapidly add PMSF to 1 mM final concentration. Remove a small amount of extract for enzyme and protein assay and for SDS–PAGE (see Part C of this exercise).

2. Slowly add enough solid ammonium sulfate (13.4 g) to make the solution 25% (w/v) in ammonium sulfate. You may wish to consult a table of fractionation with ammonium sulfate (found in most biochemistry books). Continue stirring for 10 minutes, then centrifuge at 10,000 g for 10 minutes at 4°C.

3. Decant and save the supernatant; note the volume. Dissolve the precipitate in Z buffer (about 1–2 volumes of precipitate). It is desirable to keep the protein as concentrated as possible to improve stability and prevent excessive dilution. Material that does not dissolve is probably particulate or denatured protein and can be removed and discarded after centrifugation (1000 g for 5 minutes).

4. Assay the redissolved precipitate and the supernatant for β-galactosidase activity and for protein content. Be aware that the added ammonium sulfate in the supernatant might distort the enzymatic assay.

5. Continue in this way to take precipitates in the saturation ranges of 25–40% (add an additional 8.4 g of ammonium sulfate), 40–60% (add 12.0 g of ammonium sulfate), 60–80% (add 12.9 g of ammonium sulfate), and 80% supernatant. As $(NH_4)_2SO_4$ may

Table 19.1

Saturation range (%)	β-Galactosidase precipitated (%)	Protein precipitated (%)	Purification factor[a]
0–25	?	?	?
25–40	?	?	?
40–60	?	?	?
60–80	?	?	?
80	?	?	?

[a] Specific activity of fraction/specific activity of original cell activity.

lead to acidification of the solution, add NaOH if required to keep the solution in the neutral range (pH 6.5–7.5).

6. Prepare a table, using Table 19.1 as an example, and fill in the data points (indicated by a ?).

7. Dissolve the ammonium sulfate precipitate containing the majority of enzymatic activity in Z buffer and dialyze against 1 liter of Z buffer overnight at 4°C, with one buffer change to remove excess salt. Save 100 μl of the dialyzed extract for SDS–PAGE (see Part C of this exercise).

Days 2 and 3

8. Carry out column chromatography of the ammonium sulfate-precipitated enzyme in the cold room. Apply the dialyzed sample to the gel filtration column as described in Part A of this exercise. Elute and collect fractions using the same procedures as in column calibration. Assay fractions for protein content and β-galactosidase activity. Pool and save fractions with maximum activity; store at 4°C.

PART C   SODIUM DODECYL SULFATE–POLYACRYLAMIDE GEL ELECTROPHORESIS

Introduction

Analysis of proteins by gel electrophoresis with respect to purity and molecular weight is a common procedure in most biotechnology laboratories. Modern adaptations of the technique of sodium dodecyl sulfate–polyacrylamide gel electrophoresis (SDS–PAGE) facilitate the study of such proteins in a rapid and efficient manner.

The porosity of a gel formed as a result of the copolymerization of acrylamide with the cross-linker bisacrylamide acts as a molecular sieve in which macromolecules of various sizes and charges can be separated. In the presence of the negatively charged detergent SDS and a reducing agent, commonly 2-mercaptoethanol, disulfide bonds in and between polypeptides are broken; the resulting protein subunits are bound in the form of micelles by SDS, thereby providing all molecules bound with an overall negative charge. This serves as a method by which the varying charges of the specific amino acid residues of different polypeptides are eliminated as a factor in the electrophoretic separation of these molecules, resulting in a means of differentiation based, in general, on the molecular weight of the molecules. Therefore, on application of an electric field through an SDS-containing protein separation gel, the migration of smaller, low molecular weight polypeptides will occur faster than that of larger, high molecular weight polypeptides. Depending on the size of the molecules to be separated and the relative quantities of proteins or nucleic acids to be analyzed, the percent concentration of acrylamide can be changed, as desired, to achieve optimal results.

In this part of the exercise, we will utilize SDS–PAGE to examine the protein content of cell extracts of a genetically engineered strain of *Saccharomyces cerevisiae* that has been induced to produce the enzyme β-galactosidase, which has a tetrameric molecular weight of ~460,000. The analysis of these extracts in comparison to known protein standards will provide a method by which the identification and confirmation of the β-galactosidase protein can be achieved rapidly.

Reagents/Supplies

Safety Note
Wear gloves. Acrylamide is a neurotoxin.

Acrylamide-bis [30% (w/v) acrylamide and 0.8% (w/v) bisacrylamide: dissolve 30 g of acrylamide and 0.8 g of bisacrylamide in 80 ml of double-distilled water; once dissolved, bring to 100 ml with double-distilled water, filter through Whatman No. 1 paper, and store at 4°C in a brown glass bottle].

Ammonium persulfate (100 mg/ml; prepare fresh before use)

Binder clips

Destaining solution (50 ml of methanol, 860 ml of double-distilled water, and 90 ml of glacial acetic acid; add in this order and mix well)

Extracts from *Saccharomyces cerevisiae* YNN281(pRY121)
1. Crude cell extract
2. Ammonium sulfate dialysate
3. Gel filtration fractions

β-Galactosidase standard (from *Escherichia coli,* Sigma G6008)

Gel combs

Gel spacers

Glass baking dishes (8 × 12 inches, 2 qt)

Glass plates

Hamilton syringes (10–20 μl), if available

Molecular weight standards for SDS–PAGE (Bio-Rad 161-0309)

Razor blades

Running buffer (7.2 g of glycine, 1.51 g of Tris-HCl, and 0.5 g of SDS for 500 ml in double-distilled water)

Sample buffer, 2× [0.25 M Tris-HCl (pH 6.8) with 20 ml of glycerol, 4 g of SDS, 2 mg of bromphenol blue, and 10 ml of 2-mercaptoethanol; bring to 100 ml with double-distilled water]

Separating gel buffer [1.5 M Tris-HCl (pH 8.8)]

Sodium dodecyl sulfate [SDS, 10% (w/v)]

Stacking gel buffer [0.5 M Tris-HCl (pH 6.8)]

Staining solution (1 g of Coomassie Blue, 500 ml of methanol, 100 ml of glacial acetic acid, and 400 ml of double-distilled water; add dye to methanol, stir for 1 hour, add water and acid, stir for 30 minutes, and filter through Whatman No. 1 paper)

N,N,N',N'-Tetramethylethylenediamine (TEMED)

Vacuum flask (50 ml)

Water-saturated butanol

Wax, for sealing gels

Equipment

Constant voltage power supply

Gel casting stand

Gel shaker

Mini-vertical slab gel apparatus (PAGE gel apparatuses are available from many manufacturers)

Procedure

A. Assembly of Gels

Day 1

1. Clean and dry all materials before assembly and gel casting, using soapy water and rinsing well to remove any film, etc.

2. Assemble pieces in the gel casting stand in the following order: plate, two spacers, two plates, two spacers, and so on; continue for the desired number of gels. Place the plastic plate that fits the front of the frame on last, clamp with binder clips, and seal the assembly with melted wax. This method facilitates the pouring of five or six minigels at one time.

B. Pouring of Gels

1. Prepare 10% (w/v) acrylamide separating gels by combining the following reagents from the prepared stock solutions in a 50-ml vacuum flask: 10 ml of acrylamide-bis solution (see Safety Note, p. 145), 11.8 ml of double-distilled water, 7.5 ml of separating gel buffer, and 0.3 ml of 10% (w/v) SDS.

2. Remove any excess air in the solution by degassing for 5 minutes.

3. Add 300 μl of freshly prepared ammonium persulfate and 30 μl of TEMED stock; swirl the flask to facilitate mixing without introducing any bubbles. (*Caution:* This initiates the polymerization process, so be prepared to work quickly after this step.)

4. Pour the gel solution into the casting apparatus (or use a Pasteur pipet) to a level ~2 cm from the top of the glass plate while being careful to avoid bubble formation.

5. Quickly, using a micropipettor, layer each gel, dropwise, with water-saturated butanol. This aids in the formation of a bubble-free interface on which to pour the stacking gel. Allow the gel to polymerize for about 20 minutes.

6. Once the gel is polymerized, pour off the butanol by carefully inverting the casting unit onto a paper towel.

7. Prepare the stacking gel by combining the following solutions in a 50-ml vacuum flask: 3 ml of acrylamide-bis, 5 ml of stacking gel buffer, 11.6 ml of double-distilled water, and 0.2 ml of 10% (w/v) SDS. Degas for 5 minutes.

8. Add 200 μl of ammonium persulfate and 20 μl of TEMED. Swirl and pour into the casting unit to the top of the glass plates.

9. Working quickly, carefully insert the combs into the gels. While working them in, try to remove any bubbles that may exist by gently wiggling the combs. Allow the gel to stand for about 1 hour or until polymerization is complete.

10. Gels may be removed from the casting unit by using a razor blade to remove excess wax and to separate the gels. (*Note:* Gels may be stored for 2 or 3 days in a tightly sealed container with a little water.) Rinse the gels with water to remove excess acrylamide.

C. Running of Gels

Day 2

1. Mount a gel in the gel apparatus.

2. Pour a little running buffer into the top half of the apparatus to check for leaks. If sufficiently sealed, fill both the top and bottom of the apparatus until the gel and electrodes are covered with buffer.

3. Prepare dilutions of the three extract samples (crude, ammonium sulfate, and gel filtration) to achieve samples of 10, 20, and 40 μg of protein to load onto the gel (based on the previously determined protein concentration in the extracts). Include a 1 : 1 dilution of each sample with 2× sample buffer (containing the gel dye). Also prepare a 20-μg sample of known β-galactosidase and a sample of molecular weight standards for loading. (*Note:* If prestained electrophoresis standards are available, it is not necessary to dilute them with 2× sample buffer since the progress of these standards may be visualized while running the gel.)

4. Using two hands, gently remove the comb from the stacking gel.

5. Fill the wells with the appropriate samples in the desired order (take note of the loading order) using a 10- to 20-μl Hamilton syringe or equivalent. Rinse the syringe with a little running buffer between samples to avoid cross-contamination. Clean the syringe *immediately* after use by drawing water, then 95% (v/v) ethanol, through it using a vacuum and some tubing attached to the top of the syringe barrel.

6. Connect the terminals, turn on the power supply, and set the voltage at 160 V. Apply constant voltage until the dye front is near the end of the gel.

D. Fixing and Staining of Gels

Day 2 (Continued)

1. Turn off the power supply, disconnect the terminals, and discard the buffer. Carefully remove the gel from the apparatus.

2. Remove the spacers from the sides of the gel and pry the two plates apart using a razor blade. The stacking gel may be trimmed off at this point, if desired. Place the gel in staining solution (~100 ml) in a 2-qt glass baking dish and shake on a gel shaker for 15 minutes.

3. Pour off the staining solution and destain the gel using destaining solution for 10 minutes. Change the destaining solution and repeat two or three times over 2 hours, or until the desired resolution is achieved.

References

Fischer, L. (1986). "Gel Filtration Chromatography." Elsevier, New York.

Gordon, A. H. (1975). "Electrophoresis of Proteins in Polyacrylamide and Starch Gels." Elsevier, New York.

Green, A. A., and Hughes, W. L. (1955). Protein fractionation on the basis of solubility in aqueous solutions of salts and organic solvents. *In* "Methods in Enzymology" (S. P. Colowick and N. O. Kaplan, eds.), Vol. 1, pp. 67–87. Academic Press, New York.

Hames, B. D., and Rickwood, D. (1987). "Gel Electrophoresis of Proteins: A Practical Approach." IRL Press, Oxford, England.

Pharmacia Product Guide. "Gel Filtration: Theory and Practice." Pharmacia Corporation, Uppsala, Sweden.

Questions

1. Fill in the tabulation below for your purification of β-galactosidase.

Step	Activity (units)	Protein (mg)	Specific activity (U/mg)	Yield (%)	Purification factor
Crude extract	?	?	?	100	1.00
$(NH_4)_2SO_4$ dialysate	?	?	?	?	?
Gel filtration pooled fractions	?	?	?	?	?

2. How would you verify that the band corresponding to a molecular weight of 116,000 is the monomer of β-galactosidase?

Alternative Protocols and Experiments

Exercise 6A

Isolation and Characterization of
Auxotrophic Yeast Mutants

Exercise 9A

Large-Scale Isolation of Plasmid DNA
by Column Chromatography

Exercise 12A

Colony Hybridization

Isolation and Characterization of Auxotrophic Yeast Mutants

Introduction

Saccharomyces cerevisiae is widely recognized as an ideal eukaryotic microorganism for biotechnology applications, including production of proteins, hormones, and other pharmaceutically important compounds. Yeasts have many technical advantages, such as both stable haploids and diploids, rapid growth, clonability, and ease of mutant isolation. In addition, *S. cerevisiae* has been successfully employed in all phases of genetics, such as mutagenesis, recombination, and regulation of gene action. The ability to generate mutants of desired phenotypes is an important tool for biotechnological applications, such as increasing the amount of biosynthesis of a particular compound or producing an altered enzyme of more desirable characteristics.

Because spontaneous mutation frequencies are low, yeast is usually treated with such mutagens as UV radiation, nitrous acid, ethyl methanesulfonate (EMS), diethyl sulfate, and N-methyl-N'-nitro-N-nitrosoguanidine (MNNG) in order to enhance the production of mutants. These mutagens are remarkably efficient and can induce mutations at rates between 5×10^{-4} and 10^{-2} per gene without substantial killing. Even though there are known methods to increase the proportion of mutants by killing nonmutants with nystatin or other agents, it is usually unnecessary to use selective means to obtain reasonable yields of mutants. In this experiment, auxotrophic mutants will be isolated from MNNG-treated yeast.

Auxotrophic mutants are invaluable for the elucidation of biochemical pathways as well as for the study of the relationship between enzyme structure and function. Studies on the intermediates accumulated by amino acid auxotrophs have facilitated the unraveling of biochemical pathways.

The standard treatment described here involves treatment of wild-type yeast with MNNG. After mutagenesis, the strain is diluted and plated on a complete medium at a concentration giving about 100–200 cells per petri plate. After the cells have grown into colonies, they are transferred to various media by replica plating. Auxotrophic mutants are detected by lack of growth on a minimal medium which contains glucose, potassium phosphate, ammonium sulfate, a few vitamins, salts, and trace metals. The specific requirements of auxotrophic mutants can be determined by testing the colony from the original YEPD plate on various types of synthetic media. Putative mutants are placed on a YEPD plate according to a pattern. After they are fully grown, they are transferred to plates of minimal medium containing pools of various amino acids, purines and pyrimidines, and other metabolites, each at a final concentration of about 20 μg/ml. From the pattern of growth of a particular strain, it is possible to identify the specific requirement of the mutant strain (see the tabulation below).

Pools	1	2	3	4	5
6	Adenine	Guanine	Cysteine	Methionine	Uracil
7	Histidine	Leucine	Isoleucine	Valine	Lysine
8	Phenylalanine	Tyrosine	Tryptophan	Threonine	Proline
9	Glutamate	Serine	Alanine	Aspartate	Arginine

Generally, a colony will respond on a plate containing one of the pools from 1–5 and on another plate containing one of the pools from 6–9, thus allowing for direct identification of a single growth factor requirement. For example, a colony growing on pools 1 and 7 requires histidine, and a colony growing on 3 and 8 requires tryptophan. If a colony grows on only one of the nine pools, it requires more than one of the nutrients in that pool.

Reagents/Supplies

Amino acid pools (see Introduction to this exercise and Instructor's Note below)

Dilution blanks, sterile (9.9 and 9.0 ml)

N-Methyl-N'-nitro-N-nitrosoguanidine [MNNG, 0.1 *M* in phosphate buffer (pH 7)] (*Caution: This is a powerful mutagen*)

Saccharomyces cerevisiae X2180-1A

Test tubes with 8 ml of sterile sodium thiosulfate [5% (w/v)]

Toothpicks, sterile

Velveteen pads, sterile

YEPD plates (see Exercise 5)

YNB plates containing amino acid pools (see Exercise 5)

Equipment

Incubator at 30°C

Instructor's Note

1. One day before the start of the experiment, inoculate the entire surface of a YEPD agar slant with *S. cerevisiae* X2180-1A. Incubate at 30°C overnight.

2. On the day of class, there should be ~7.5×10^8 cells on the slant. Harvest and suspend the cells in 1.5 ml of sterile sodium phosphate buffer (pH 7.0) and add 0.7 ml of the cell suspension to 1 ml of buffer. Add 50 μl of MNNG and agitate.

3. Prepare amino acid pools (see Appendix 10 and Introduction to this exercise). Add to YNB medium so that the final concentration in YNB plates is 20 μg/ml for each amino acid, purine, or pyrimidine.

Procedure

Day 1

1. Shake the cell suspension containing MNNG at 30°C for 1 hour. Remove 100 μl of cells and add to 100 μl of sodium thiosulfate to inactivate the MNNG. Determine the cell density with a counting chamber while the cells are incubating (see Appendix 14). The mutagenesis will cause ~40% killing.

2. Transfer 0.2 ml of the treated cell suspension to a test tube containing 8 ml of sterile 5% (w/v) sodium thiosulfate, which will quench the reaction.

3. To obtain 100–200 viable cells per plate, dilute the treated cells, which are in the sodium thiosulfate solution, by a factor of 10^{-4} with sterile water (two sequential 1 : 100 dilutions can be made by twice pipetting 0.1 ml of the previous dilution into 9.9 ml of sterile water or another appropriate dilution depending on the number of cells in the original suspension and accounting for the dilutions made during the experiment).

4. Spread either 0.1, 0.2, or 0.4 ml each on separate YEPD plates, using 10 plates for each of the different volumes plated. Incubate all of the plates at room temperature (23°C) for 3 or 4 days.

Day 3 or Day 4

5. Choose 10 or more YEPD plates containing 25–200 colonies per plate for the isolation of auxotrophic mutants. Transfer the colonies from each of the YEPD plates by replica plating, using sterile velveteen pads, to one YNB plate and one YEPD plate. Be certain that each plate is numbered and has an orientation symbol on the back. For detection of auxotrophic mutants, incubate the YNB plates and the YEPD replicas at 30°C for 1 day.

Day 5

6. Compare each of the 10 minimal plates with the YEPD plates. Transfer colonies whose replicates have failed to grow on the minimal plates with toothpicks to a YEPD plate in a pattern. There should be ~5% of such colonies. Incubate at 30°C overnight.

Day 6

7. Replica plate the master plate containing the auxotrophic mutants to the nine petri plates having the various pools of metabolites. Incubate at 30°C overnight.

Day 7

8. Record the growth response and assign a number for each mutant.

References

Maga, J. A., and McEntee, K. (1985). Response of *S. cerevisiae* to *N*-methyl-*N'*-nitro-*N*-nitrosoguanidine: Mutagenesis, survival and *DDR* gene expression. *Mol. Gen. Genet.* **200**, 313–321.

Snow, R. (1966). An enrichment method for auxotrophic yeast mutants using the antibiotic "nystatin." *Nature (London)* **211**, 206–207.

Exercise 9A

Large-Scale Isolation of Plasmid DNA by Column Chromatography

Introduction

In practice, clones of transformed bacteria putatively containing a plasmid of interest can be identified by restriction digestion of DNA from mini-preps. After identification, large-scale isolation and purification of plasmid DNA are the next steps for recovering milligram amounts of plasmid. Large-scale isolation of plasmid DNA entails three basic steps: (1) growing bacteria and amplifying the plasmid, (2) harvesting and lysing the bacteria, and (3) purifying the plasmid DNA by separating host proteins and chromosomal DNA.

Selective inhibition of host chromosomal DNA replication by the addition of chloramphenicol during the log phase of growth will result in "amplification" of pRY121 (see Exercise 10, Figure 10.1). Most plasmids commonly utilized for cloning contain a "relaxed" origin of replication. This characteristic signifies that the plasmid DNA does not require cellular factors for replication, therefore the plasmid can continue replicating even after growth of the host has been arrested. Hundreds of copies of plasmid per cell can be obtained by amplification.

In the second step of isolation, the amplified bacterial cells are harvested by centrifugation. The cells are then treated with lysozyme, in the presence of sucrose for osmotic stabilization, to weaken their cell wall structure. Cells are finally lysed gently by mixing with a mild detergent—alkali solution. This facilitates the release of the compact, supercoiled molecules of plasmid into solution, whereas most of the larger chromosomal DNA will remain associated with cellular debris.

Because it exists as covalently closed circles, the double-stranded plasmid DNA is not denatured by the alkali treatment.

The final stage of isolation involves the utilization of a relatively new method—spun column chromatography—for the separation of plasmid DNA from the remaining chromosomal DNA. Spun column chromatography is rapidly becoming the technique of choice in molecular biology laboratories owing to its ease of application, low expense, and time efficiency. Column chromatography effectively acts as a substitute for standard cesium chloride density-gradient ultracentrifugation for the isolation of plasmid DNA without the labor-intensive preparation, exposure to toxic chemicals (EtBr), and expense of the latter process. One manufacturer of these columns is 5 Prime → 3 Prime, which produces the pZ523 columns used in this exercise. However, similar products from other manufacturers are also available. Plasmid DNA is collected as the effluent from centrifugation of alkaline-lysed DNA through the column while the chromosomal DNA remains bound to the column and is discarded. The resulting purified DNA is ready for additional purification, analysis, or manipulation.

Reagents/Supplies

Ammonium acetate (10 M)
Ampicillin (25 mg/ml)
Chloramphenicol
Chloroform–isoamyl alcohol [24 : 1 (v/v)]
Corex centrifuge tubes, sterile (15 ml)
Escherichia coli LE392(pRY121)
Ice-cold ethanol (100%)
Isopropanol (100%)
LB broth (see Exercise 3)
Lysozyme (prepare fresh; 5 mg/ml of Solution A below)
NaCl–TE solution (prepare 4 M NaCl in TE buffer)
Phenol–chloroform–isoamyl alcohol [25 : 24 : 1 (v/v)]

pZ523 buffers

Prepare 100 ml of the following solutions

Solution A: 50 mM glucose, 25 mM Tris-HCl (pH 8.0), and 10 mM EDTA

Solution B: 0.2 N NaOH and 1% (w/v) SDS; filter sterilize, prepare fresh

Solution C: 60 ml of 5 M potassium acetate, 11.5 ml of acetic acid, and 28.5 ml of distilled water; store at 4°C

pZ523 columns

RNase (DNase-free; see Exercise 7)

TE buffer (pH 8.0; see Exercise 7)

Equipment

Beckman J2-21 centrifuge (or equivalent) at 4°C

Beckman JA-10 rotor (or equivalent)

Beckman JA-20 rotor (or equivalent)

Beckman TJ-6 tabletop centrifuge with swinging buckets (or equivalent)

Freezer at −20°C

Freezer at −70°C

Microcentrifuge

Shaking incubator or water bath at 37°C

Spectrophotometer

Instructor's Note

1. Two days prior to the start of the experiment, inoculate 5 ml of LB broth containing 50 μg/ml of ampicillin with *E. coli* LE392(pRY121). Incubate overnight with shaking at 37°C.

2. One day prior to the start of the experiment, inoculate 1000 ml of LB broth containing 50 μg/ml of ampicillin in a 2-liter flask with 2 ml of the overnight *E. coli* LE392(pRY121) culture. Incubate with shaking at 37°C.

3. When the culture reaches a density (OD$_{600}$) of 0.7–0.8, amplify the plasmid DNA in the culture by adding 200 mg of chloramphenicol (solid); swirl the flask well to mix the chloramphenicol. Incubate overnight at 37°C with shaking.

Procedure

Day 1

1. Harvest the *E. coli* cells by centrifugation in two sterile 500-ml polycarbonate bottles. Spin at 8000 *g* using a Beckman JA-10 rotor at 4°C for 10 minutes.

2. Discard supernatant and freeze the pellets for 10 minutes at −70°C. (This step is optional.)

3. Thaw pellets on ice and resuspend one of the pellets in 10 ml of Solution A containing 5 mg/ml of freshly prepared lysozyme. Transfer the suspension to the other tube and resuspend the remaining pellet. Split into two equal volumes in 40-ml sterile polycarbonate centrifuge tubes. Incubate at room temperature for 5 minutes.

4. Add 20 ml of Solution B to each tube. Mix by inversion (*do not vortex*). Incubate on ice for 10 minutes.

5. Add 15 ml of ice-cold Solution C to each tube. Mix by inversion (*do not vortex*).

6. Centrifuge at 10,000 *g* in a Beckman JA-20 rotor for 20 minutes at 4°C.

7. Recover the supernatant, carefully, and transfer to two 40-ml polycarbonate centrifuge tubes. Record the volume of supernatant.

8. Add 0.6 volumes of 100% isopropanol, mix, and allow to stand at room temperature for 20 minutes.

9. Centrifuge at 10,000 *g* for 25 minutes at 4°C. Discard the supernatant.

10. Resuspend one pellet in 2 ml of TE buffer (pH 8.0) and transfer the suspension to the remaining pellet and thoroughly resuspend.

11. Add RNase (DNase-free) to a final concentration of 100 μg/ml. Incubate at room temperature for 20 minutes.

12. Add 10 M ammonium acetate to a final concentration of 2 M; mix by vortexing. Transfer the mixture to a 15-ml Corex centrifuge tube.

13. Extract once with an equal volume of phenol–chloroform–isoamyl alcohol.

14. Centrifuge the mixture in a tabletop centrifuge (Beckman TJ-6) at 4000 g for 4 minutes. Remove the upper aqueous phase and transfer to a clean 15-ml Corex tube.

15. Extract this phase with an equal volume of chloroform–isoamyl alcohol. Centrifuge and transfer the upper aqueous phase to a clean screw-cap 15-ml polypropylene tube.

16. Add 2 volumes of ice-cold 100% ethanol. Mix by inversion. Store at −20°C until the next laboratory period.

Day 2

1. Transfer the solution to a polycarbonate centrifuge tube and recover the plasmid DNA by centrifugation at 10,000 g in a Beckman JA-20 rotor for 25 minutes at 4°C.

2. Resuspend the pellet in 1.35 ml of TE buffer and 0.45 ml of NaCl–TE solution. Place the DNA solution on ice.

3. Remove the top and then the bottom closures of the pZ523 column and allow excess equilibration buffer to drain from the column.

4. Place the column in one of the collection tubes provided with the pZ523 column and spin the assembly for 1 minute at 1100 g in a tabletop centrifuge with swinging buckets (e.g., 1000 rpm in a Beckman TJ-6 centrifuge).

5. Discard the collected effluent and tube. It is normal to notice that the resin has been packed down ∼1 cm below the column support after this spin.

6. Very carefully apply 1.8 ml of the sample solution from Step 2 onto the column bed using a micropipettor.

7. Centrifuge the column–collection tube assembly at 1100 *g* for 24 minutes.

8. Recover the column effluent containing the purified plasmid DNA and transfer to two or three microcentrifuge tubes. Note the volume in each tube.

9. To each microcentrifuge tube, add 10 *M* ammonium acetate to a final concentration of 2 *M*.

10. Add 0.6 volumes of 100% isopropanol, mix, and allow to stand at room temperature for 20 minutes.

11. Centrifuge for 30 minutes at room temperature at top speed in a microcentrifuge. Discard the supernatant.

12. Invert the tubes on clean absorbent paper and allow the pellets to dry for 10–15 minutes.

13. Plasmid DNA should be resuspended and combined in a total volume of 0.5–1.0 ml of TE buffer and stored at 4°C.

Reference

Zervos, P. H., Morris, L. M., and Hellwig, R. J. (1988). A novel method for rapid isolation of plasmid DNA. *BioTechniques* **6**, 238–242.

Colony Hybridization

Introduction

Colony hybridization allows microbial clones or virus plaques to be screened by nucleic acid hybridization to indicate those clones containing a sequence of interest (Grunstein and Hogness, 1975). Colonies are transferred to filters and lysed *in situ* to denature and immobilize the DNA. The transferred DNA can then be probed by a specific cloned sequence, synthetic oligonucleotide, or RNA to determine whether the host cell contained nucleic acid that hybridized to the probe.

In this exercise, clones of *Escherichia coli* and *Saccharomyces cerevisiae* will be screened to determine the presence of β-galactosidase sequences. The procedure will use the same biotinylated probe as in Exercise 12.

Reagents/Supplies

Agar plates (LB and LB plus ampicillin; see Exercise 14)
Agar plates (YEPD; see Exercise 4)
Biotinylated probe and indicator system (see Exercise 12)
Escherichia coli LE392
Escherichia coli LE392(pRY121)
Nitrocellulose paper or nylon membrane
Plastic wrap
Saccharomyces cerevisiae YNN281
Saccharomyces cerevisiae YNN281(pRY121) (see Exercise 14)
Sodium hydroxide (NaOH, 0.5 *M*)
Tris-HCl [1 *M* (pH 7.4)]

Tris-HCl [1 *M* (pH 7.4)] containing NaCl (1.5 *M*)
Whatman 3 MM paper

Equipment

Incubator at 30°C
Incubator at 37°C
Vacuum oven at 80°C

*Instructor's Note*_____

1. Grow fresh streak plates of *S. cerevisiae* YNN281 and *S. cerevisiae* YNN281(pRY121) on YEPD at 30°C. Grow fresh streak plates of *E. coli* of LE392 and *E. coli* LE392(pRY121) on LB and LB plus ampicillin, respectively, at 37°C.
2. Use sterile toothpicks to transfer colonies onto plates containing YEPD (*S. cerevisiae*) or LB with or without ampicillin (*E. coli*) medium. Make 2- to 3-mm streaks (small patches) in a grid pattern of about 25 colonies.
3. Incubate inverted plates at 37°C (*E. coli*) or 30°C (*S. cerevisiae*) for 1–2 days.

Procedure

Day 1

1. Place nitrocellulose or nylon filters cut to fit petri dishes onto the agar plates. Handle filters with forceps. Let stand 1–2 minutes.

2. Mark the filter and master plate in three locations with a pencil. The master plate should be sealed with Parafilm and stored at 4°C.

3. Remove the filter from the plate by peeling off carefully and float *colony side up* in 0.5 *M* NaOH. (About 1 ml of 0.5 *M* NaOH on plastic wrap is convenient.) Be sure the filter is wet evenly and leave at room temperature for 3 minutes.

4. Blot the side of the filter exposed to NaOH onto a paper towel. Repeat Step 3 with fresh 0.5 M NaOH.

5. Transfer the filter to 1 ml of 1 M Tris-HCl (pH 7.4). Let stand for 5 minutes to permit neutralization, blot, and repeat.

6. Transfer to 1 ml of 0.5 M Tris-HCl (pH 7.4) containing 1.5 M NaCl. Let stand for 5 minutes, blot, transfer to dry Whatman 3 MM paper, and let dry at room temperature for 45 minutes.

7. Place the filter between two sheets of Whatman 3 MM paper and bake at 80°C in a vacuum oven for 90 minutes.

8. Remove the filter and store at 4°C until the next laboratory period.

Day 2

9. Hybridize probe as in Part D of Exercise 12.

References

Grunstein, M., and Hogness, D. S. (1975). Colony hybridization: A method for the isolation of clone DNAs that contain a specific gene. *Proc. Natl. Acad. Sci. U.S.A.* **72**, 3961–3966.

Hanahan, D., and Meselson, M. (1983). Plasmid screening at high density. *In* "Methods in Enzymology" (R. Wu, L. Grossman, and K. Moldave, eds.), Vol. 100, pp. 333–342. Academic Press, New York.

2

Buffer Solutions

A buffer is a solution containing a mixture of a weak acid (HA) and its conjugate base (A^-) that is capable of resisting substantial changes in pH on addition of small amounts of acidic or basic substances. Addition of acid to a buffer leads to conversion of some of the A^- to the HA form; addition of base leads to conversion of some of the HA to the A^- form. As a result, addition of either acid or base leads to a change in the A^-/HA ratio. In the Henderson–Hasselbalch equation, which applies to buffer systems, the ratio A^-/HA appears as a logarithmic function. Hence, changes in the A^-/HA ratio lead to only minor changes in pH within the working range of the buffer.

The working range of a buffer is determined by the pK value of the system. Specifically, it falls within 1 pH unit of the pK value. Beyond the working range, not enough of both of the buffer forms is present to allow the buffer to function effectively when either acid or base is added. Buffer capacity depends on the volume of the buffer and on the concentrations of the two buffer components. Buffer capacity is usually defined as the number of equivalents of either H^+ or OH^- required to change the pH of a given volume of buffer by 1 pH unit.

A buffer may be prepared in two ways: (1) known amounts of the A^- and HA forms may be mixed and diluted to volume or (2) to a known amount of the HA form (A^- form), a known amount of base (acid) may be added and the mixture diluted to volume. Note that the molarity of a buffer always refers to the total concentration of the buffer species. Thus, a 0.5 M $H_2PO_4^-$/HPO_4^{2-} buffer is one in which the sum of the concentrations of $H_2PO_4^-$ and HPO_4^{2-} is 0.5 moles per liter.

Henderson–Hasselbalch Equation

The Henderson–Hasselbalch equation is derived from the dissociation (ionization) reaction for a weak acid (HA):

$$HA = H^+ + A^-$$

where HA is a Brönsted acid (proton donor, conjugate acid) and A is a Brönsted base (proton acceptor, conjugate base). The two species HA and A are said to form a conjugate acid–base pair. The equilibrium (dissociation, ionization) constant for this reaction (K or K_{eq}) is given by

$$K = \frac{[H^+][A^-]}{[HA]}$$

where brackets indicate molar concentration (moles per liter). By taking logarithms and rearranging, the Henderson–Hasselbalch equation is obtained

$$pH = pK + \log \frac{[A^-]}{[HA]}$$

where pH is equal to the negative logarithm of the hydrogen ion concentration ($-\log [H^+]$) and $pK = -\log K$.

As indicated, the pH is defined as

$$pH = -\log [H^+] = \log \frac{1}{[H^+]}$$

This definition (attributed to Arrhenius) was chosen in order to obtain small positive numbers for the hydrogen ion concentration. The pH scale usually runs from 0 to 14, with 7 representing neutrality; pH values above 7 characterize basic conditions and pH values below 7 represent acidic conditions. A change of 1 unit in pH represents a 10-fold change in the hydrogen ion concentration.

In pure water, and in dilute solutions, the product of the hydrogen ion and hydroxide ion concentrations (known as the ion product of water, K_w) is a constant:

$$[H^+][OH^{-1}] = K_w = 10^{-14} \qquad \text{(at 25°C)}$$

3

Preparation of Buffers and Solutions

When making up solutions, the following guidelines should be observed.

1. Use the highest grade of reagents, when possible.

2. Prepare all solutions with the highest quality distilled water available.

3. Autoclave solutions when possible. If the solution cannot be autoclaved and you wish to store it, sterilize by filtration through a 0.22-μm filter.

4. Check the pH meter carefully, using freshly prepared solutions of standard pH, before adjusting the pH of buffers.

5. Always label containers with the name of the solution, percentage or concentration, your name/course, and date. For highly basic solutions, such as 1 M NaOH, be sure to use plastic containers as glass is corroded by bases.

6. Store solutions cold when possible.

Time can be saved by using concentrated stock solutions (e.g., 1 M Tris) to make up a range of different solutions.

Percentage Solution

A percentage solution is one in which the exact concentration of the solute is known in 100 ml of a liquid or solution. The concentration may be expressed or determined by weight or volume; it may be written out in grams and milliliters and expressed as follows:

Percentage (w/v) = weight (g) in 100 ml of solution
Percentage (v/v) = volume (ml) in 100 ml of solution

Examples by Weight

For a 1% (w/v) aqueous solution of sodium chloride, weigh out 1 g of NaCl and add to 100 ml of distilled water, or 10 mg of NaCl per milliliter.

For a 5% (w/v) aqueous solution of sodium chloride, weigh out 5 g of NaCl and add to 100 ml of distilled water.

For 1000 ml of a 5% (w/v) aqueous solution of sodium chloride, use 10 times the above amounts, that is, 50 g of NaCl in 1000 ml of distilled water.

Example by Volume

The concentration of a solution is stated on the label, and a percentage solution should be determined from this value. For example, commercial strength hydrochloric acid varies from 36 to 40% (v/v). If you want a 1% (v/v) solution of HCl, it is erroneous to take 1 ml of 36–40% (v/v) HCl and add it to 99 ml of distilled water.

An acceptable and correct formula for determining the amount of 36% (v/v) HCl needed to make up 100 ml of a 1% (v/v) aqueous solution of HCl is the following:

Starting concentration (%) $\times$ unknown volume (ml)
= final concentration (%) $\times$ total volume (ml)
$$36 \times X = 1 \times 100$$
$$36X = 100$$
$$X = \frac{100}{36}$$
$$X = 2.78 \text{ ml of 36\% (v/v) HCl in 97.22 ml of}$$
distilled water

See Appendix 4 for common commercial strengths of acids and bases.

Dilutions of Solutions (Percentage by Volume)

The most accurate formula for making dilutions of solutions is the following:

$$\text{Percentage you have} \times \text{unknown volume (ml)}$$
$$= \text{percentage you want} \times \text{volume wanted (ml)}$$

Example of Preferred Method

To make up 1000 ml of 70% (v/v) ethanol from 95% (v/v) ethanol, substitute the known quantities in the above formula:

$$95 \times X = 70 \times 1000$$
$$95X = 70{,}000$$
$$X = \frac{70{,}000}{95}$$
$$X = 736.8 \text{ ml of 95\% (v/v) ethanol in 263.2 ml of}$$
$$\text{distilled water}$$

Although the above method is preferred, another generally accepted method is subtracting the percentage required from the percentage strength of the solution that is to be diluted. The difference will be the amount of water that is to be used.

Example of Subtraction Method

The percentage required is 70% (v/v) ethanol, the solution to be diluted is 95% (v/v) ethanol. As $95 - 70 = 25$, 70 parts of 95% (v/v) alcohol and 25 parts of distilled water are the amounts to be combined. In order to make up ~1 liter of 70% (v/v) alcohol, you would simply place 700 ml of 95% (v/v) alcohol in a 1000-ml graduated cylinder and fill to the 950 ml mark with distilled water.

Molar Solutions

A molar (mole, molecular) solution is one in which 1 liter (1000 ml) of the solution contains the number of grams of the solute equal to its molecular weight (sum of atomic weights).

Example of Molar Solutions

To make up a 1 M solution of sodium chloride.

1. Obtain the atomic weights of the elements (see a Periodic Table).

    Sodium (Na) = 22.997
    Chlorine (Cl) = 35.459

2. Obtain the sum of atomic weights from the formula for molecular weight.

    Molecular weight of NaCl = 22.997 + 35.459 = 58.456

3. Weigh 58.46 g of NaCl and make up to a total volume of 1000 ml with distilled water.

To make up 100 ml of a 1 M solution of sodium chloride.

1. Same as Step 1 above.

2. Same as Step 2 above.

3. $\dfrac{58.46}{10} = 5.85.$

4. Weigh 5.85 g of NaCl and make up to a total volume of 100 ml with distilled water.

To make up a M/10 (1/10 molar or 0.1 M) solution of sodium chloride.

1. Same as Step 1 above.

2. Same as Step 2 above.

3. $\dfrac{58.46}{10} = 5.85.$

4. Weigh 5.85 g of NaCl and make up to a total volume of 1000 ml with distilled water.

4

Properties of Some Common Concentrated Acids and Bases

Table A-4.1

Reagent[a]	Molecular weight	Approximate specific gravity of reagent	Molarity of concentrated reagent	Amount of reagent (ml) to make 1 liter of a 1 M solution
Hydrochloric acid (HCl)	36.46	1.19	12.1	82.6
Nitric acid (HNO₃)	63.02	1.42	15.8	63.3
Hydrofluoric acid (HF)	20.01	1.19	29.0	34.5
Perchloric acid (HClO₄)	100.46	1.67	11.7	85.5
Glacial acetic acid (H₃CCO₂H)	60.05	1.05	17.4	57.5
Sulfuric acid (H₂SO₄)	98.08	1.84	18.0	55.5
Phosphoric acid (H₃PO₄)	98.00	1.70	14.8	67.6
Ammonium hydroxide (NH₄OH)	35.05	0.90	14.8	67.6
Sodium hydroxide (NaOH)	40.00	1.53	19.3	51.8
Potassium hydroxide (KOH)	56.11	1.46	11.7	85.5

[a] Be sure to check the labels of the reagents used in your laboratory as the molarity and specific gravity vary with manufacturer.

5

Use of Micropipettors

Many companies manufacture continuously adjustable and fixed pipets to deliver volumes from 1 to 5000 μl. The micropipettors are generally factory calibrated. However, a careful check of accuracy and reproducibility of volume delivery should be made to ensure correct usage. The operating directions described below are for the Rainin Pipetman as suggested by the manufacturer. For the micropipettor used in your laboratory, be sure to read the instruction manual provided by the manufacturer.

Operation

1. Set the desired volume by holding the Pipetman body in one hand and turning the volume adjustment knob until the correct volume shows on the digital indicator. The friction O ring of the volume adjustment locks the volume securely. For the best volume setting precision, always approach the desired volume by dialing downward (at least one-third revolution) from a larger volume setting.

2. Attach a new disposable tip to the shaft of the pipet. (A major source of error in the use of micropipettors is the variation between sources of disposable tips. It may be desirable to compare experimentally tips from one supplier with those of another.

Refer to Exercise 2 for a procedure to determine the standard deviation and coefficient of variation of tips.) Press tip on firmly with a slight twisting motion to ensure a positive, airtight seal.

> Models P-20, P-100, and P-200 use RC-20 yellow tips
> Model P-1000 uses RC-200 blue tips
> Model P-5000 uses C-5000 white tips

3. Depress the plunger to the *first positive stop*. This part of the stroke is the calibrated volume displayed on the digital indicator.

4. Holding the Pipetman vertically, immerse the disposable tip into the sample liquid to a depth of

> 1–2 mm for RC-20 tips (P-20, P-100, P-200)
> 2–4 mm for RC-200 tips (P-1000)
> 3–6 mm for C-5000 tips (P-5000)

5. Allow the push button to return *slowly* to the up position. *Never permit it to snap up.*

6. Wait 1 or 2 seconds to ensure that the full volume of sample is drawn into the tip.

7. Withdraw the tip from the sample liquid. Any liquid remaining on the outside of the tip should be wiped carefully with a lint-free cloth, taking care not to touch the tip opening.

8. To *dispense the sample,* place the tip end against the side wall of the receiving vessel and depress the plunger slowly to the *first stop*. Wait (longer for viscous solutions) for

> 1 second for RC-20 (yellow) tips
> 1–2 seconds for RC-200 (blue) tips
> 2–3 seconds for C-5000 (white) tips

Then depress the plunger to the *second stop* (bottom of stroke) to expel any residual liquid in the tip.

9. With the plunger fully depressed, withdraw the Pipetman from the vessel carefully by allowing the top to slide along the wall of the vessel.

10. Allow the plunger to return to the *top position.*

11. Discard the tip by depressing the tip ejector button. A fresh tip should be fitted for each sample to prevent carry-over between samples.

Pipetting Guidelines and Precautions

Consistency in all aspects of the pipetting procedure will contribute significantly to optimum reproducibility. Therefore, attention should be given to

1. *Speed and smoothness* during depression and release of the push button

2. *Pressure* on the push button at the first stop

3. *Immersion depth*

4. *Minimal angle* from the vertical axis

If an **air bubble** is noted within the tip during intake, dispense the sample into the original vessel, check the tip immersion depth, and pipet more slowly. If an air bubble appears a second time, discard the tip and use a new one.

6

Safe Handling of Microorganisms

Many of the procedures used in molecular biology research involve the use of live microorganisms. Whenever such organisms are used, it is essential that laboratory workers adhere rigidly to a microbiology laboratory code of practice and thereby significantly reduce the possibility of a laboratory-acquired infection. The safest way to approach work with live microorganisms is to make the following assumptions.

1. Every microorganism used in the laboratory is potentially hazardous.

2. Every culture fluid contains potentially pathogenic organisms.

3. Every culture fluid contains potentially toxic substances.

The basis of a microbiology laboratory code of practice is that no direct contact should be made with the experimental organisms or culture fluids, for example, contact with the skin, nose, eyes, or mouth. It must also be noted that a large proportion of laboratory-acquired infections result from the inhalation of infectious aerosols released during laboratory procedures. Below is a list of rules that forms the basis of a microbiology laboratory code of practice.

1. A laboratory coat that covers the trunk to the neck must be worn at all times.

2. There must be no eating, drinking, or smoking in the laboratory.

3. There must be no licking of adhesive labels.

4. Touching the face, eyes, and so on should be avoided.

5. There must be no chewing or biting of pens or pencils.

6. Available bench space must be kept clear, clean, tidy, and free of unessential items, such as books and handbags.

7. No materials should be removed from the laboratory without express permission of the laboratory supervisor or safety officer.

8. All manipulations, such as by pipet or loop, should be performed in a manner likely to prevent the production of an aerosol of the contaminated material.

9. Pipetting by mouth of *any* liquid is strictly forbidden. Pipet fillers or automatic pipets are used instead.

10. All manipulations should be performed aseptically, using plugged, sterile pipets, and the contaminated pipets should be immediately sterilized by total immersion in a suitable disinfectant.

11. Contaminated glassware and discarded petri dishes must be placed in lidded receptacles provided for their disposal.

12. All used microscope slides should be placed in receptacles containing disinfectant.

13. It must be recognized that certain procedures or equipment, for example, agitation of fluids in flasks, produce aerosols of contaminated materials. Lids should be kept on contaminated vessels, when possible.

14. All accidents, including minor cuts, abrasions, and spills of culture fluids or reagents, must be reported to the laboratory supervisor or safety officer.

15. Before leaving, swab your working area with an appropriate disinfectant fluid.

16. Whenever you leave the laboratory, wash your hands with a germicidal soap and dry them with paper towels. Laboratory coats should be removed and stored for future use or laundered. Do not under any circumstances wander into an office or restroom area wearing a potentially contaminated laboratory coat.

Appendix _____

7

List of Cultures

Table A-7.1

Exercise	Organism[a]	Strain designation	Source
3	*Escherichia coli*	—	Any strain from stock culture collection, such as ATCC,[b] or available at your institution
4	*Escherichia coli*	—	As in Exercise 3
	Saccharomyces cerevisiae	—	As in Exercise 3
6	*Saccharomyces cerevisiae*	YNN281[c]	YGSC[d]
7	*Escherichia coli*	LE392(pRY121)[e]	ATCC 37658
14	*Escherichia coli*	LE392[f]	ATCC 33572
17	*Saccharomyces cerevisiae*	YNN281(pRY121)[g]	Created by students in Exercise 13
6A	*Saccharomyces cerevisiae*	X2180-1A	YGSC[d]

[a] Genotypes of organisms used are listed in the respective exercises under Reagents/Supplies.

[b] American Type Culture Collection (12301 Parklawn Drive, Rockville, Maryland 20852).

[c] Also used in Exercises 12A, 13, and 17.

[d] Yeast Genetics Stock Center (Department of Biophysics, University of California, Berkeley, Berkeley, California 94720).

[e] Also used in Exercises 9, 9A, and 12A.

[f] Also used in Exercise 12A.

[g] Also used in Exercises 12A, 18, and 19 (Part B).

Appendix _____

8

Storage of Cultures

Numerous methods are used to store strains. A permanent but time-consuming method for *Escherichia coli* is freeze-drying (lyophilization); freeze-dried strains have been kept alive over 30 years. Another permanent and useful method for *E. coli* and *Saccharomyces cerevisiae* is ultracold storage at $-70°C$ or in liquid nitrogen. To prevent bursting of the cells during freezing, the culture medium must contain either 7% (w/v) dimethyl sulfoxide (DMSO) or 15% (w/v) glycerol. To recover cells from frozen cultures, the stored culture is scraped with a sterile loop or sterile toothpick, and the ice crystals are allowed to thaw on an agar plate (and then streaked) or placed in sterile growth medium. If the glycerol content is increased to 40–50% (w/v), cultures can be stored in a liquid state for several years at $-20°C$.

For convenience, stable strains of *E. coli* can be stored at room temperature as a deep stab culture in airtight tubes containing rich agar. Stab cultures last several years. They are prepared by autoclaving a screw-cap tube containing LB agar and by allowing the agar to harden. Bacteria are picked with a wire loop from an agar plate or a liquid culture, and the loop is stabbed into the rich agar. After the tube is placed in an incubator and the bacteria are grown for 24 hours, the tube is sealed with paraffin and stored at room temperature. The lifetime of a deep stab is determined primarily by the quality of the airtight seal, because it is essential that the agar does not dry.

For routine use in the laboratory, a slant culture is used for

E. coli or *S. cerevisiae*. Medium is placed in a small screw-cap tube and sterilized. The tube is tipped to increase the surface area, and the agar is allowed to harden. The agar surface is then covered with bacteria or yeast either by using a loopful of cells or by placing a droplet of culture directly on the surface. The tube is incubated overnight to allow for cell growth, and the cap is then tightened. Slant cultures are stored in a refrigerator and will last several months, less if opened frequently.

9

Sterilization Methods

Table A-9.1

Method	Time and temperature	Use	Mode of action	Disadvantages
Dry heat	One hour at 170°C in a hot-air sterilizer	Glassware, pipets, petri dishes, metals, forceps, etc.	Oxidizes bacterial components	Cannot be used for media sterilization because of high temperatures; most media not stable
Intermittent sterilization ("Tyndal-lization")	Thirty minutes in flowing steam (100°C) for 3 consecutive days	Media, tissues, liquids, solids	Heat shock the material for the first 30 minutes, and if spores are present, they will germinate and the vegetative cell will be destroyed at the next cycle	Tedious, time-consuming

(continued)

Table A-9.1 *Continued*

Method	Time and temperature	Use	Mode of action	Disadvantages
Moist heat	Fifteen to thirty minutes at 121°C under 15 psi pressure	Glassware, media, liquids, solids, metals, etc.	Coagulates proteins of micro-organisms	Some nutrients are not stable under these conditions, e.g., lactose
Filtration	Need Millipore filter, vacuum, and Millipore apparatus	Media, liquids	Millipore filters contain pores so small that bacteria cannot pass through	Glassware cannot be sterilized this way
Toxic gases (cold sterilization)	Ethylene oxide under slight pressure	Large surface areas, pieces of equipment, hospital rooms, surgery suites, etc.	Kills all life forms; lethal gas	Flammable; sometimes hard to remove all traces
Radiation	Ultraviolet (UV) or ionizing (γ or X-rays)	Media, liquids, solids, pharma-ceuticals	Causes genetic mutations, interferes with metabolism and respiration, and leads to death	Does not penetrate glassware easily; requires access to a radioactive core or cathode ray tube

Appendix

10

Preparation of Stock Solutions for Culture Media

Table A-10.1

Component	Dissolve in	Sterilization[a]
Amino acids (% w/v)		
L-Alanine (0.5)	H_2O	Autoclave
L-Arginine (2)	H_2O	Autoclave
L-Asparagine (0.5)	H_2O	Autoclave
DL-Aspartic acid (2)	2 N NaOH	Filter
L-Cysteine (0.5)	H_2O	Filter
L-Glutamic acid (0.5)	2 N NaOH	Filter
Glycine (2)	H_2O	Autoclave
L-Histidine (1)	H_2O	Autoclave
L-Isoleucine (0.5)	H_2O	Autoclave
L-Leucine (2)	H_2O	Autoclave
L-Lysine (1)	H_2O	Autoclave
DL-Methionine (1)	H_2O	Autoclave
DL-Phenylalanine (1)	H_2O	Autoclave
L-Proline (2)	H_2O	Autoclave
DL-Serine (2)	H_2O	Autoclave
DL-Threonine (1)	H_2O	Autoclave
DL-Tryptophan (0.5)	1 N HCl	Filter
L-Tyrosine (0.5)	2 N NaOH	Filter
DL-Valine (2)	H_2O	Autoclave

(*continued*)

Table A-10.1 *Continued*

Component	Dissolve in	Sterilization[a]
Purines and pyrimidines (% w/v)		
Adenine (0.5)	0.1 N HCl	Filter
Cytosine (0.5)	H_2O	Autoclave
Guanine (0.5)	0.1 N HCl	Filter
Thymine (0.1)	H_2O	Autoclave
Uracil (0.2)	H_2O	Autoclave
Xanthine (0.5)	0.1 N NH_4OH	Filter

[a] For filtration use a 0.45- or 0.2-μm membrane filter.

11

Growth in Liquid Medium

Growth Curve

Cells growing in a batch (shake flask) culture generally experience four distinct growth phases: a lag phase, an exponential (log) phase, a stationary phase, and a death phase. Cells in the lag phase, such as an aliquot from an older culture that has been transferred to fresh medium, do not grow right away. The cells must adjust to the new medium before growth will begin at a rapid rate. The length of the lag phase is dependent on a number of factors: age and genotype of the inoculum, temperature, nutrient levels of both the old and new media, aeration, and the concentration of toxins that may have been formed in the old medium. The lag phase of *Saccharomyces cerevisiae* at 30°C in YEPD lasts for about 3 hours. For *Escherichia coli* at 37°C, the lag phase is usually 10–60 minutes in LB.

Once the cells begin growing rapidly, they are said to enter the exponential, or logarithmic (log), growth phase. Cells in the log phase are growing rapidly, and, unlike cells in the lag and stationary phases, most are in the same physiological state. The growth rate during log phase is dependent on the nutrient level and aeration of the medium. Oxygen is frequently the limiting factor in yeast cultures, and the cultures must be shaken rapidly for sufficient oxygen to dissolve. The doubling time of *S. cerevisiae* grown in YEPD at 30°C may be between 90 and 100 minutes, while a 200-minute doubling time may be expected in minimal medium. The presence of an introduced plasmid may slow growth since plasmid replication requires energy and metabolites. At 37°C, a typical *E. coli* strain doubles in 20–30 minutes in rich medium and in 50–60 minutes in minimal medium. The growth rate constant (μ) serves to define the rate of growth of a culture during balanced growth:

$$\mu = \ln 2 \div g$$

where *g* is the mean doubling time or generation time. For further discussion of this concept and derivation of the formula, see Mandelstam *et al.* (1982).

As nutrients within the flask are consumed and inhibitory products accumulate, the growth rate slows and eventually stops while the culture enters the stationary phase. Cells in a stationary phase culture are not all in the same physiological state; some are dividing while others are dying. Only the overall population size remains constant. As more nutrients are depleted, more cells die than are produced, and the culture enters the death phase.

Aeration

Escherichia coli and *S. cerevisiae* can grow both aerobically and anaerobically, although anaerobic growth does not occur with all carbon sources. With glucose, anaerobic growth occurs but at about 10% the rate of aerobic growth. At a cell density of $\sim 10^7$ cells/ml in liquid medium, oxygen cannot diffuse from the atmosphere to the cells fast enough for aerobic growth, and, unless the medium is shaken or directly aerated by bubbling, growth of a culture slows considerably at such densities. Even with aeration, the oxygen supply is inadequate above a cell concentration of $2-3 \times 10^9$ cells/ml for *E. coli* or $2-3 \times 10^8$ cells/ml for *S. cerevisiae*.

Inoculation and Subculture

Since many cells in an inoculum obtained from a slant culture are dead, most experiments begin by inoculating a small volume of growth medium and then growing the cells overnight. In the morning, cell growth will have stopped, and the cell density will usually be $2-3 \times 10^9$ cells/ml for *E. coli* in rich medium and $2-4 \times 10^8$ cells/ml for *S. cerevisiae* in rich medium. To obtain actively growing cells for any particular experiment, the overnight culture is diluted 10–100 times and is regrown for several hours, at which time the cells are usually in the exponential growth phase. Also, if an exponentially growing culture is rapidly chilled to below 8°C (by shaking a growth flask in ice water), is stored at 4°C, and then is rewarmed rapidly to the original temperature at which the culture had been growing, growth will resume without a lag.

Reference

Mandelstam, J., McQuillen, K., and Dawes, I. (1982). "Biochemistry of Bacterial Growth," 3rd Ed. Halsted, Oxford, England.

12

Determination of Viable Cells

One cell can multiply until a single visible colony forms. This is the basis of counting cells by plating because a count of the number of colonies produced by a particular volume of a culture indicates the number of viable cells in the culture. In an exponentially growing culture or in a culture that is in early stationary phase, all cells can usually form colonies. Thus, the colony count approximately equals the number of viable cells.

To obtain a reliable count by plating, it is essential that single colonies not be formed by two or more cells. Accordingly, the number of cells placed on an agar surface should not exceed a few hundred, and the cells should be spread evenly. Since the concentration of cultures used in laboratory experiments usually exceeds 10^6 cells/ml, the culture must be diluted prior to plating. The standard procedure is to make a series of sequential 10- to 100-fold dilutions until the cell concentration is a few thousand cells per milliliter. Then, a 0.1-ml sample is spread on an agar surface; this operation (dilution and spreading) is called plating. To avoid errors that might arise by transferring very small volumes, a volume of 0.05–0.5 ml is usually transferred. We use 10- and 100-fold dilutions prepared, respectively, by adding 0.5 ml of cells to 4.5 ml of sterile diluent and 0.05 ml of cells to 4.95 ml of sterile diluent. (Many laboratories use 1 to 9.0 and 0.1 to

9.9.) Each dilution must be done with a separate pipet, otherwise cells remaining in an earlier dilution may be carried over to a later dilution tube. A major cause of dilution errors is the transfer of liquid on the outside of the pipet. This could be avoided by carefully wiping the pipet, which is usually not possible, however, because of resulting contamination. Therefore, to minimize errors, it is best to submerge the tip as little below the surface of the liquid as possible, touch the side of the tube to remove any adhering droplets, and then blow out the liquid into the diluent. The necessity to blow out the liquid is the reason that serological (blowout) pipets are used for dilutions rather than analytical (to deliver) pipets.

When the final aliquot of 0.1 ml of diluted cells is placed on the agar, the droplet is spread over the surface with a glass spreader (hockey stick). The spreader is sterilized by dipping it in ethanol, shaking off the excess liquid, and igniting the remaining alcohol in the flame from a Bunsen burner. The spreader is cooled by touching the agar surface and then used to spread the droplet uniformly over the surface. If liquid remains on the agar, cells will drift through the liquid and, after cell division, two colonies might form from one cell initially deposited. Plates prepared a day in advance usually absorb the liquid rapidly; absorption is accelerated by spreading the droplet as thinly as possible. After the surface of the plate is dry, the plate is placed in an incubator. As the plate warms, liquid is exuded from the uppermost surface of the agar. Thus, it is advisable to invert the plate to avoid puddles on the surface of the agar that cells could drift through. This also prevents droplets, which might condense on the inside surface of the top plate, from falling on the agar.

13

Determination of Cell Mass

Growth of cultures can be monitored by spectrophotometry. The number of photons scattered is proportional to the mass of the cells in a sample (except for very concentrated cultures, as discussed below) or, for particular growth conditions, to the cell concentration. However, the geometry of each spectrophotometer determines how much scattered light falls onto a photodetector. Therefore, a calibration curve that relates cell concentration and absorbance must be made for each spectrophotometer. This is done by measuring the absorbance of suspensions of cells at different concentrations and concomitantly determining the cell concentration by direct counting or by plating on agar and measuring the number of colonies formed.

It is important to note that the absorbance is a measure of cell mass rather than cell number. Cell size varies with growth phase, so it is best to calibrate the spectrophotometer with exponentially growing cells because such cells are used in most experiments. Cell size also varies from one growth medium to the next, decreasing as the medium becomes poorer, so a calibration curve is needed for each growth medium, and often for each strain. As a rough guide, for many *Escherichia coli* strains $A_{600} = 1.0$ for $\sim 8 \times 10^8$ cells/ml and for *Saccharomyces cerevisiae* $A_{600} = 1.0$ for $\sim 3 \times 10^7$ cells/ml.

If the cell density is too high, a photon may be deflected away from the photodetector by one cell and then back again by a second cell. This effect causes the absorbance to be lower than if multiple scattering were not occurring; it becomes important at A_{600} values above 0.7. Thus, when the concentration of a dense culture is to be determined by spectrophotometry, the culture is diluted prior to reading. The measured value is then corrected by the dilution factor.

Wavelengths other than 600 nm can be employed in determining cell density, and, in fact, the sensitivity increases as the wavelength decreases. Wavelengths as short as 400 nm may be used, but not with all rich media. Rich media usually absorb short-wavelength light significantly, which complicates measurements.

14

Determination of Cell Number

To find the total number of cells in a culture, direct microscopic counting is the method of choice. For yeast cells, a hemocytometer can be used. Magnification of $100\times$ to $400\times$ is required to visualize the counting chamber and the suspended cells. For bacterial cells, a Petroff-Hauser chamber and $1000\times$ magnification are required because of the small cell size. Direct counting is often advantageous because of the opportunity afforded to visualize directly the morphology of the cells studied. Aberrant or unusual morphology could indicate suboptimal growth conditions or the presence of an altered cell genotype or phenotype.

Charging the Chamber

Vortex the cell suspension. Withdraw some of the suspension into a Pasteur pipet. Deposit a small drop on the polished surface of the counting chamber next to the edge of the cover glass. The suspension will enter the chamber by capillary action. In a properly filled chamber, cells fill only the space between the cover glass and counting chamber. No fluid should run down into the moat.

Performing the Count

Place the hemocytometer on the stage of a compound microscope and focus on the grid lines under low magnification. For yeast counting, use the middle square (see Figure A-14.1). (In Figure A-14.1, the middle square is circled and contains 25 squares, each bounded by three grid lines indicated in the figure by heavy black lines. Each of the 25 squares contains 16 smaller squares.) Adjust the light intensity so that both the cells and the lines are plainly visible. Depending on cell numbers, count all the cells in the middle square or in the five squares indicated by an X within the middle square (Figure A-14.1). Include

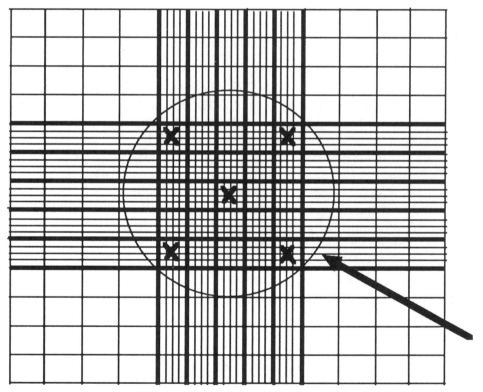

Figure A-14.1   Hemocytometer chamber. The middle encircled area is seen under 100-fold magnification.

those cells which touch the boundary lines but do not overlap the boundaries into other areas.

Calculations

The middle square has a volume of 0.1 mm^3. The number of cells in the middle square multiplied by 10 will yield the number of cells in 1.0 mm^3 in the culture. That number of cells multiplied by 1000 will yield the number of cells in 1.0 cm^3 or in 1.0 ml of the culture. Therefore, if the total middle square is counted, multiply that number by 10^4 to yield the cells per milliliter. Alternatively, if only five squares within the middle square are counted, the cell number should be multiplied by 5 × 10^4 to yield the cells per milliliter. Whichever volume is chosen, a total of more than 100 cells should be counted to obtain an accurate cell count.

15

Nomenclature of Strains

The standard nomenclature for genes and genetic markers for bacteria is that of Demerec *et al.* (1966). A three-letter italicized or underlined symbol is used to describe a genetic locus, for example, *his* (histidine) or *gal* (galactose). All histidine-requiring mutants are written *his* or *his⁻*. Often there are several enzymes in a locus; for instance, several enzymes are needed to synthesize histidine, and a deficiency of any of these will be *his⁻*. A capital letter following the gene designation is used to distinguish loci, for example, *hisA* and *hisB*. Usually, each locus represents a particular polypeptide chain or regulatory element (operator or promoter) in the system. Each independently isolated mutation within a particular locus is given a number, for example, *hisA38*. Not all *hisA* mutations are identical or yield exactly the same phenotype, but all *hisA38* mutations are identical. The phenotype of a cell is written with a capital letter and is not italicized; a haploid cell carrying the *hisA38* marker has the His⁻ phenotype, whereas a wild-type histidine prototroph is designated His⁺.

The genotype of a bacterial strain is usually given by listing all loci known to be different from the wild type. When necessary, a + is appended to a genetic symbol to indicate a wild-type allele of a locus; for example; *hisC⁺* signifies a wild-type gene.

For *Saccharomyces cerevisiae,* the following are examples of genetic nomenclature:

URA3	Locus or dominant allele
ura3	Locus or recessive allele that results in the requirement of uracil
URA3$^+$	Wild-type allele
ura3-52	Specific allele or mutation at the *URA3* locus
Ura$^+$	Phenotypic designation of a strain not requiring uracil
Ura$^-$	Phenotypic designation of a strain requiring uracil

The gene symbols differ somewhat from those for bacteria in that the genetic locus is identified by a number rather than a letter following the three-letter gene symbol. Recessive alleles are symbolized by lowercase italicized letters, whereas dominant alleles are represented by uppercase italicized letters. Different alleles for the same locus are differentiated by a number separated by a hyphen and following the locus number. There are a number of exceptions to these general rules. For example, the wild-type alleles of the mating type locus are designated *Mat*a and *Mat*α, and mutations are written *mat*a-*1* and *mat*α-*1*. The phenotypes corresponding to these alleles are a and α, respectively. For further discussion of genetic nomenclature, see Sherman (1981) and Sherman *et al.* (1986).

A detailed linkage map of *Escherichia coli* has been published by Bachmann (1983). The article by Bachmann also contains a valuable list of references to original research papers describing mutations found in many strains; moreover, it contains useful references to techniques used in modern microbial genetics. Most strains of *E. coli* that you might want can be obtained free of charge from Dr. Barbara Bachmann (Department of Human Genetics, Yale University School of Medicine, 333 Cedar Street, New Haven, Connecticut 06510). A detailed linkage map of *S. cerevisiae* has been published as well (Mortimer and Schild, 1985). Strains of *S. cerevisiae* can be purchased for a minimal charge from the Yeast Genetics Stock Center (Department of Biophysics, University of California, Berkeley, Berkeley, California 94720). The American Type Culture Collection (12301 Parklawn Drive, Rockville, Maryland 20852) is also an excellent source for *E. coli* and *S. cerevisiae* strains.

References

Bachmann, B. (1983). Linkage map of *Escherichia coli* K-12, ed. 7. *Microbiol. Rev.* **47**, 180–230.

Demerec, M., Adelberg, E. A., Clark, A. J., and Hartman, P. E. (1966). A proposal for a uniform nomenclature in bacterial genetics. *Genetics* **54**, 61–76.

Mortimer, R. K., and Schild, D. (1985). Genetic map of *Saccharomyces cerevisiae*, ed. 9. *Microbiol. Rev.* **49**, 181–213.

Sherman, F. (1981). Genetic nomenclature. *In* "Molecular Biology of the Yeast *Saccharomyces*" (J. N. Strathern, E. Jones, and J. Broach, eds.), pp. 639–640. Cold Spring Harbor Laboratory, Cold Spring Harbor, New York.

Sherman, F., Fink, G. R., and Hicks, J. B. (1986). "Laboratory Course Manual for Methods in Yeast Genetics." Cold Spring Harbor Laboratory, Cold Spring Harbor, New York.

16

Glassware and Plasticware

Experiments in molecular biology require that all items of glassware and other equipment are scrupulously clean before use. This is particularly important for reactions involving proteins and nucleic acids, which are usually carried out in small volumes, because dirty test tubes, bacterial contamination, or traces of detergent can easily inhibit reactions or accidentally degrade nucleic acids by the action of unwanted nucleases. Following thorough washing (dichromate–sulfuric acid solution is very effective for cleaning glass pipets), glass or plastic vessels and pipets should be thoroughly rinsed in distilled or double-distilled water and sterilized by autoclaving. Alternatively, heat-resistant glassware can be baked in an oven at 150°C for 1 hour.

Glassware can be treated with 0.1% (v/v) diethyl pyrocarbonate overnight at 37°C as a precaution against ribonuclease contamination. The diethyl pyrocarbonate is removed by heating the drained glassware at 100°C for 15 minutes. This procedure is not normally required for sterile disposable plasticware.

Very small quantities of nucleic acids can be absorbed onto the surface of glassware. The problem is significantly reduced if small polypropylene (e.g., microcentrifuge) test tubes are used; if these are not available, however, you may wish to coat glassware with silicone. To do this, soak or rinse glassware in a 5% (v/v) solution of dichlorodimethylsilane in chloroform. This solution can be stored at room temperature in a fume hood virtually indefinitely and used as required. Following the silicone treatment, thoroughly rinse the glassware and bake at 180°C overnight.

Many experiments with nucleic acids involve the use of solutions containing phenol and/or chloroform. Whereas these solutions can be successfully used in glass or polypropylene (microcentrifuge) tubes, they attack other forms of plastic, used in many types of centrifuge tubes, particularly polycarbonate. Check the manufacturer's recommendations if you are doubtful about using a particular type of plastic tube and remember that sealing films are also soluble in phenol and chloroform.

17

Preparation of Tris and EDTA

1 *M* Tris Stock

Weigh out 121.1 g of Tris base [tris(hydroxymethyl)aminomethane; MW 121.1 g/mole] and place into a 2000-ml beaker containing a magnetic stir bar. Add 800 ml of distilled water and dissolve the Tris with stirring. Adjust the pH to the desired value by addition of concentrated hydrochloric acid (HCl). The addition of HCl should be done carefully, while the solution is stirring and being monitored for pH, to avoid a drastic drop in the pH. The amount of concentrated HCl to be added differs according to the desired pH of the Tris solution:

pH 7.4: ~70 ml of HCl
pH 7.6: ~60 ml of HCl
pH 7.8: ~50 ml of HCl
pH 8.0: ~42 ml of HCl

Check the pH, pour the solution into a graduated cylinder (or volumetric flask), and bring up the volume to 1000 ml using distilled water. Mix and dispense as necessary and sterilize by autoclaving. Tris stocks of different molarities are prepared in the same manner.

Table A-17.1   Preparation of 50 m*M* Tris Buffer

pH at 25°C	Tris-HCl (g/liter)	Tris base (g/liter)
7.4	6.61	0.97
7.6	6.06	1.39
7.8	5.32	1.97
8.0	4.44	2.65

Notes

1. It is desirable to obtain a calomel electrode, which is an electrode compatible with Tris that can be purchased from most manufacturers. With silver pH electrodes, Tris precipitates, causing clogging of the electrode tip.

2. The pH of Tris buffers varies greatly with temperature. For example, a pH of 8.0 at 25°C becomes 8.58 at 5°C and 7.71 at 37°C. For the exercises in this course, the pH of Tris at 25°C (room temperature) is used even if the solutions are cooled or heated. For some experiments, however, the pH variation with temperature must be taken into account.

3. A method for the preparation of a 50 mM Tris stock solution if Tris base and Tris-HCl are available involves simply mixing these two compounds as indicated in Table A-17.1.

0.5 *M* EDTA Stock

Weigh out 186.1 g of disodium EDTA (disodium ethylenediaminetetraacetic acid, dihydrate; MW 372.2 g/mole) and place into a 2000-ml beaker. Add 800 ml of distilled water and stir vigorously, using a stir bar and a magnetic stirrer. Adjust the pH to 8.0 using concentrated sodium hydroxide (NaOH) to facilitate dissolution of the EDTA. This can be accomplished by addition of concentrated

NaOH liquid using a Pasteur pipet or by addition of ~20 g of NaOH pellets while mixing and monitoring the pH of the EDTA solution. The solution is then brought to 1000 ml in a graduated cylinder (or volumetric flask), is mixed, and is dispensed into aliquots for sterilization by autoclaving.

18

Basic Rules for Handling Enzymes

Note _____
This is a reprint of the Boehringer Mannheim Biochemicals primer which was first published in the December 1985 issue of *BM Biochemica*

For the novice: Basic hints to guide you through your first enzymatic reaction.

For the expert: A refresher course and an aid for training your students.

1. For best stability, enzymes should be stored in their original commercial form (lyophilized, ammonium sulfate suspension, etc.), undiluted, and at the appropriate temperature as specified on the label.

2. For enzyme solutions and assay buffers, use the highest purity water available. Glass-distilled water is best. Deionized water, especially if passed through an old deionizing filter or a reverse osmosis device, may contain traces of organic contaminants which inhibit enzymes.

3. Enzymes should be handled in the cold (0–4°C). Dilute for use with ice-cold buffer or distilled water, as appropriate for each enzyme. While using the enzyme solution or suspension at the bench, keep it in an ice bath or ice bucket.

4. Dilute enzyme solutions are generally unstable. The amount of enzyme required for the experiment should be diluted within 1–2 hours of use. Enzymes should not be diluted for long-term storage. Enzymes, especially those which have been diluted, should be checked for activity periodically to ensure that any slight loss in activity is taken into account when designing an experimental protocol.

5. Do not shake crystalline suspension (e.g., ammonium sulfate suspension) since oxygen tends to denature the enzyme. The material should be resuspended with gentle swirling or by rolling the bottle on the laboratory bench. Once the enzyme crystals have been uniformly resuspended, remove the amount needed with a pipet. In many cases, the enzyme crystals may be used directly in assay procedures.

6. Do not freeze crystalline suspensions. Freezing and thawing in the presence of high salt concentrations cause denaturation and loss of activity.

7. Vials containing lyophilized enzymes (as well as cofactors, such as NADH and NADPH) should be warmed to room temperature before opening. This prevents condensation of moisture onto the powder, which can cause loss of activity or degradation. If the reagent is hygroscopic, one such mishandling may well ruin the entire vial.

8. Avoid repeated freeze-thawing of dilute enzymes and lyophilisates in solution. Store in small aliquots. Thaw one portion at a time and store that portion, once thawed, at 4°C. The stability of individual enzymes may vary greatly and often should be determined empirically under your exact conditions.

9. Detergents and preservatives should be used with caution, since

they may affect enzyme activity. Sodium azide, for example, inhibits many enzymes which contain heme groups (e.g., peroxidase). Detergents added at concentrations above their critical micellar concentration form micelles which may entrap and denature the enzyme.

10. Enzymes should be handled carefully to avoid contamination of any kind. Use a fresh pipet for each aliquot that is removed from the parent vial. Never return unused materials to the parent vial. Wear gloves to prevent contaminating the enzyme with proteases, DNases, RNases, and inhibitors often found on fingertips. Never pipet by mouth.

11. Adjust the pH of the enzyme buffer at the temperature at which it will be used. Many common buffers (Tris, glycylglycine, Bes, Aces, Tes, Bicine, HEPES) change pH rapidly as the temperature changes. For instance, Tris buffer decreases 0.3 pH units for every 10°C rise in temperature. A solution of Tris, adjusted to pH 7.5 at 25°C, will have a pH of 8.1 at 4°C or 7.2 at 37°C. The change in pH per 10°C temperature change for other buffers is as follows: Aces, −0.20; Bes, −0.16; Bicine, −0.18; glycylglycine, −0.28; HEPES, −0.14; Tes, −0.20 (Good *et al.*, 1966).

12. The absorbance at 280 nm, widely used to quickly determine the protein concentration of an enzyme solution, actually is due to the presence of tyrosine and tryptophan in the protein. If an enzyme (e.g., superoxide dismutase) contains a low amount of these two amino acids, it will not absorb significantly at 280 nm.

Detailed information is available on many enzymes. The most complete references are as follows.

Methods in Enzymology, published by Academic Press, Editors-in-chief: John N. Abelson and Melvin I. Simon. To date, there are 185 volumes published in this series, covering an extensive range of topics.

The Enzymes, 3rd Edition, published by Academic Press, edited by Paul D. Boyer. This excellent, broad series focuses more on the physical and biological properties of enzymes and less on methodology than *Methods in Enzymology.*

Methods of Enzymatic Analysis, 3rd Edition, published by Verlag Chemie, Editor-in-chief: Hans U. Bergmeyer. This book gives in-depth discussions of analysis techniques which use enzymes or which assay enzymes.

Reference

Good, N. E., Winget, G. D., Winter, W., Connolly, T. N., Izawa, S., and Singh, R. M. M. (1966). Hydrogen ion buffers for biological research. *Biochemistry* **5**, 467–477.

19

Manufacturers' and Distributors' Addresses

This appendix is designed to assist the instructor in obtaining the reagents, equipment, and supplies needed for teaching this course. For a complete listing of biotechnology products and instruments, see *Guide to Scientific Instruments* [*Science* **243** (February 24, 1989, Part 2)]. This guide is published yearly by the American Association for the Advancement of Science.

Major Distributors of General Laboratory Equipment and Supplies

Baxter Scientific Products (General Offices)
1430 Waukegan Road
McGaw Park, Illinois 60085
(312) 689-8410

Beckman Instruments, Inc.
Spinco Division
1117 California Avenue
Palo Alto, California 94340
(415) 857-1150

DuPont (Sorvall)
Biotechnology Systems
Wilmington, Delaware 19818
(800) 551-2121

Fisher Scientific (varies with location)
Headquarters:
711 Forbes Avenue
Pittsburgh, Pennsylvania 15219
(412) 562-8300

PGC Scientifics
9161 Industrial Court
Gaithersburg, Maryland 20877
(301) 840-1111

Thomas Scientific
99 High Hill Road
P.O. Box 99
Swedesboro, New Jersey 08085
(800) 524-0027

Chromatography

Bio-Rad Laboratories
1414 Harbour Way South
Richmond, California 94804
(800) 227-3259 West
(800) 645-3227 East

Pharmacia LKB Biotechnology, Inc.
800 Centennial Avenue
Piscataway, New Jersey 08854
(800) 558-7110

5 Prime → 3 Prime, Inc.
19 East Central Avenue
Paoli, Pennsylvania 19301
(215) 644-4710

Electrophoresis Apparatuses and Power Supplies

Bethesda Research Laboratories, Inc. (BRL)
P.O. Box 6009
Gaithersburg, Maryland 20877
(800) 638-8992

Bio-Rad Laboratories
(see Chromatography, p. 211, for address)

Hoefer Scientific Instruments
654 Minnesota Street
Box 77387
San Francisco, California 94107
(800) 227-4750

Idea Scientific Company
P.O. Box 2078
Corvallis, Oregon 97339
(503) 758-0999

Fine Chemicals

Aldrich Chemical Company, Inc.
940 West Saint Paul Avenue
Milwaukee, Wisconsin 53223
(800) 558-9160
(414) 273-3850

Amersham Corporation
2636 South Clearbrook Drive
Arlington Heights, Illinois 60005
(800) 323-9750

Calbiochem Corporation
P.O. Box 12087
San Diego, California 92112
(800) 854-3417

FMC Bioproducts
5 Maple Street
Rockland, Maine 04840
(800) 341-1475
(207) 594-3200

J. T. Baker Chemical Company
22 Red School Lane
Phillipsburg, New Jersey 08865
(201) 859-5411

Sigma Chemical Company
P.O. Box 14508
St. Louis, Missouri 63178
(800) 325-3010

Microbiological Media

BBL Microbiology Systems
P.O. Box 243
Becton Dickinson Co.
Cockeysville, Maryland 21030
(800) 638-6663

Difco Laboratories
P.O. Box 1058
Detroit, Michigan 48232
(313) 961-0800

Also see Major Distributors of General Laboratory Equipment and
Supplies, pp. 210–211.

Microbiological Strains

American Type Culture Collection (ATCC)
12301 Parklawn Drive
Rockville, Maryland 20852
(800) 638-6597

Yeast Genetics Stock Center (YGSC)
Department of Biophysics
University of California, Berkeley
Berkeley, California 94720
(415) 642-0815

Restriction Enzymes and Reagents for Molecular Biology

Bethesda Research Laboratories, Inc. (BRL)
(see Electrophoresis Apparatuses and Power Supplies, p. 212, for
address)

Boehringer Mannheim Biochemicals (BMB)
9115 Hague Road
P.O. Box 50816
Indianapolis, Indiana 46250
(317) 576-2771

International Biotechnologies Inc. (IBI)
275 Winchester Avenue
P.O. Box 9558
New Haven, Connecticut 06535
(800) 243-2555
(203) 562-3878

New England Biolabs, Inc. (NEB)
32 Tozer Road
Beverly, Massachusetts 01915
(800) 632-5227
(617) 927-5054

Promega Corporation
2800 South Fish Hatchery Road
Madison, Wisconsin 53711
(800) 356-9526

Sigma Chemical Company
(see Fine Chemicals, p. 213, for address)

Stratagene Cloning Systems
11099 North Torrey Pines Road
La Jolla, California 92037
(800) 548-1113

United States Biochemical Corporation
P.O. Box 22400
Cleveland, Ohio 44122
(800) 321-9322

Glossary

The terms are defined in the context of this course. Commonly used jargon is also defined.

absorbance The absorption of part of the visible spectrum by an object or solution. The quantitative relationship of absorbance as a function of the length of the light path and the concentration of the absorbing species is the principle behind spectrophotometry.

agar An extract of red algae (family Rhodophyceae) used as a solidifying agent for microbiological media.

alkaline phosphatase An enzyme that can be conjugated to biotin to function as part of the detection system for biotinylated probes. Reaction of the enzyme with the substrate BCIP (5-bromo-4-chloro-3-indolyl phosphate) generates a colored product.

ampicillin A semisynthetic penicillin that inhibits cell wall synthesis in bacteria. It is commonly used in molecular biology for selection of ampicillin-resistant microorganisms.

amplification of plasmid To increase the copy number of a relaxed plasmid that continues to replicate despite conditions of arrested growth and division of the host cell.

anode The positive terminal, usually colored red in a gel electrophoresis apparatus. Negatively charged nucleic acid molecules migrate to the anode from the cathode when an electrical field is applied.

aseptic technique A set of standard, commonsense rules of microbiological practice which ensures that procedures are executed with a minimum of risk to the worker and without contamination of samples and the work space.

aspirate To remove a liquid layer, such as a supernatant, from a sample using a pipet or equivalent attached to a vacuum source with a trap to catch effluent.

biotinylation of nucleic acids A nonradioactive method of labeling nucleic acid probes using nick translation to incorporate biotin-derivitized nucleotides. An alternative to radioactive labeling.

bromphenol blue A chemical dye used as a pH indicator which progressively turns from yellow to blue over the range of pH 3–5. Also commonly used as a tracking dye for nucleic acids in electrophoresis gels. The dye migrates at ~200 base pairs.

bromthymol blue   A chemical dye used as a pH indicator which progressively turns from yellow to blue over the range of pH 6–7.5.

buffer   A conjugate acid–base pair which functions as a system for resisting changes in pH; especially important for maintaining pH during studies of biological systems *in vitro*.

bugs   (slang)   Microorganisms, usually bacteria.

cathode   The negative terminal, usually colored black in a gel electrophoresis apparatus. Negatively charged DNA molecules will migrate from the cathode to the anode when an electrical field is applied.

cesium chloride (CsCl)   A heavy inorganic salt used for creating density gradients in a high-speed ultracentrifuge to purify nucleic acids.

chloramphenicol   A broad-spectrum antibiotic from *Streptomyces venezuelae* which interferes with peptide bond formation in prokaryotes. Used as a selection agent and for amplifying relaxed plasmids.

clone   (1) A genetically identical population of cells descended from one cell. (2) To purify a genetically identical population of cells from a large number of genetically heterogeneous cells. May also refer to a gene or piece of DNA.

colony   A visible growth of microorganisms or cells on solid microbiological media. A colony may or may not be a clone.

comb   A piece of plastic or Teflon shaped to make wells or slots of specific dimensions in an electrophoresis gel. Samples are loaded into the wells.

competent   Refers to a particular condition of cells, such as *E. coli* following chemical treatment, to make the cell envelope permeable to exogenous DNA molecules.

complex medium   Nutritional medium for microorganisms which includes ingredients, such as yeast extract and bacto-peptone, for which the exact chemical composition is not known.

cuvette   A small plastic or quartz vessel of specific dimensions and light-absorbing qualities designed to hold a sample for spectrophotometry.

defined medium   Nutritional medium for microorganisms for which the exact chemical composition is known.

denaturation of nucleic acids   To cause strand separation of double-stranded nucleic acid by heating or treating with alkali to enable hybridization of the resulting single-stranded species, such as a labeled probe, to another single-stranded nucleic acid target.

dialyze   To remove impurities, such as salts, from a solution of macromolecules by allowing diffusion of smaller molecules across a semipermeable membrane into water or an appropriate buffer.

DNA polymerase I The *E. coli* DNA polymerase enzyme which contains an excision repair function that is taken advantage of in nick translation reactions to incorporate labeled nucleotides into a DNA probe.

electrophoresis A method used to separate charged molecules which migrate in response to the application of an electrical field. In gel electrophoresis, heterogeneously sized molecules in a sample are drawn through an inert matrix, such as agarose, and separate during migration according to size.

3′ end The end of a piece of DNA (or RNA) that contains the phosphate group attached to the 3′ carbon of the corresponding base.

5′ end The end of a piece of DNA (or RNA) that contains the hydroxyl group attached to the 5′ carbon of the corresponding base of the nucleotide.

ethidium bromide (EtBr) An orange dye (and strong mutagen) that intercalates the stacked bases of nucleic acids (double-stranded most efficiently) and causes the molecules to fluoresce under UV light. Fluorescence allows visualization of fragments in gels and cesium chloride density gradients and also provides a way to quantitate nucleic acids whereby the intensity of fluorescence is compared to known standards.

filter sterilize To sterilize a liquid medium or medium component, in opposition to autoclaving, by suction through a selective membrane into a sterile container.

flame To sterilize a utensil used in microbiology, such as a loop or needle, by inserting it into a Bunsen burner flame until it is red hot. Also refers to briefly passing the mouths of glass bottles through the flame after use to minimize the risk of contaminating the contents. A part of aseptic technique.

hockey stick (slang) A piece of solid glass rod about 15 cm long and 5 mm in diameter that has been bent into the shape of a hockey stick. It is used to spread small volumes of solutions or cells on microbiological plates after flame sterilizing the bent end with ethanol.

host The organism that has been infected or transformed with a virus or plasmid and that is intended to be a growth or study "chamber" for the newly introduced DNA on the viral or plasmid vector.

hybridize To allow complementary strands of DNA, RNA, or DNA and RNA to anneal to yield a double-stranded product. In practice, *in vitro* hybridization of labeled DNA to DNA immobilized on a Southern blot locates homologous sequences under specific salt and temperature conditions.

kilobase (kb) A unit of length for nucleic acid strands equal to 1000 bases or nucleotides.

Klett Name of a colorimeter model which uses colored filters and light scattering to measure the density of a cell culture in arbitrary "Klett units."

λ *Hind*III   A set of DNA fragments of 0.5–23.1 kb generated by digesting the phage λ with the restriction enzyme *Hind*III. The fragments are used as standard size markers on agarose gels and are commercially available.

β-**lactamase**   A bacterial enzyme, encoded by the *bla* gene, that cleaves the β-lactam–thiazolidine ring of penicillin antibiotics. Organisms containing the *bla* gene are ampicillin resistant. *bla* is commonly included on plasmid constructions as a selectable marker (usually designated *Amp*^r) for transformed hosts, such as *E. coli*.

LB (**Luria–Bertani**)   A complex rich medium for culturing bacteria.

log phase   Stage of exponential microbial growth, following the lag phase, when cells are dividing at a constant rate.

loop   A utensil used in microbiology consisting of a straight handle with a wire ~4 inches long inserted into it. The wire can be flame sterilized, and the end is twisted into a loop which is convenient for inoculating cultures, streaking plates, etc.

lysozyme   An enzyme that removes the bacterial cell wall and causes cell lysis in the presence of EDTA. Used in plasmid DNA isolation and purification procedures.

M9   A defined minimal medium for culturing bacteria.

map   (1) To determine the linear order of restriction sites on a piece of DNA by performing a series of digests and analyzing the fragment banding pattern on a gel. (2) The order of the restriction sites generated by mapping.

master plate   A plate that contains the original microbial colonies from which replica plates were made.

microcentrifuge   A small tabletop centrifuge with a radius of 40–50 mm and rotor holes for microcentrifuge tubes which is capable of speeds of up to 14,000 rpm.

mini-prep   A rapid 1- or 2-day procedure for small-scale isolation and purification of plasmid DNA from an *E. coli* host which entails growth and lysis of the bacteria, differential centrifugation in a microcentrifuge, and purification by solvent extraction.

nick   A small gap in one strand of double-stranded DNA caused by mechanical stress, UV light, or an enzyme, such as DNase I.

nick translation   An *in vitro* method for labeling DNA. *Escherichia coli* DNA polymerase I can use a nick as a starting point to excise nucleotides progressively and replace them with labeled nucleotides. The nick progresses or is "translated" as the polymerase moves 5′ to 3′ along the DNA template until another nick is encountered. ("Translation," as used here, has nothing to do with RNA translation into protein.)

nitrocellulose   A transfer medium on which nucleic acids or proteins are immobilized by blotting.

nylon membrane   A transfer medium for blotting DNA or protein purported to be stronger and more versatile than nitrocellulose.

overnight   (slang) A broth culture of bacteria that had been inoculated the previous day and allowed to grow overnight into the stationary phase to be used as a source of bacteria the next day; likely to be subcultured.

plasmid   A self-replicating circle of extrachromosomal DNA. Plasmids are used in molecular biology as cloning vectors to introduce foreign DNA into a host cell. Plasmids occur naturally in bacteria and usually impart some selective advantage to the host, such as antibiotic resistance.

plate   (1) A petri dish. (2) To spread cells onto solid nutrient medium in a petri dish.

polyethylene glycol (PEG)   An organic reagent used to alter the cell envelope of yeast to facilitate transformation.

probe   (1) A piece of labeled DNA or RNA used to locate immobilized sequences on a blot by hybridizing under optimal conditions of salt and temperature. (2) To hybridize a probe to a blot.

quadrant streak   A style of streaking for isolated single colonies of microorganisms on solid medium which entails dividing the plate into four areas or quadrants.

refractile fish   (slang) Undissolved agarose particles remaining after insufficiently heating an agarose–buffer mixture to solubilize the agarose.

restrict   To digest with a restriction enzyme.

restriction enzyme   An enzyme that cleaves double-stranded DNA at specific recognition sequences of nucleotides. Used extensively in molecular biology in standard procedures, such as mapping or modifying DNA for cloning.

restriction site   The recognition sequence for a restriction enzyme. May specifically refer to the exact point of cleavage between nucleotides within the sequence.

selectable marker   A gene that permits survival of a selected phenotype. Plasmids usually contain selectable markers, such as antibiotic resistance genes, for identification of transformants from an antibiotic-sensitive background of cells cultured on a selective medium.

Southern blot   (1) To transfer DNA electrophoretically or more commonly by capillary action from a gel to an appropriate medium, such as nitrocellulose or a nylon membrane. (2) The piece of transfer medium containing the DNA sequences that have been blotted.

sterilization   The process of eradicating all life forms. Culture media, glassware, and utensils are sterilized before use to prevent contamination. Accomplished by irradiation, heat, etc.

stop buffer   A buffer containing a component, such as the metal-chelating agent EDTA, that will stop an enzymatic reaction. Added to restriction digests at the end of incubation or sometimes included in the tracking dye solution.

streak   To use a loop, toothpick, or other sterile utensil to dilute microorganisms on solid medium in an effort to obtain isolated colonies.

streptavidin   A component of the "Dagwood sandwich" detection system for biotinylated probes. It is capable of binding to both probe and alkaline phosphatase because it is tetravalent.

subculture   To reinoculate fresh culture medium with cells from an existing culture, such as from an overnight or streak plate.

T_m   The melting temperature of a nucleic acid hybrid which is related to G + C content and also to the temperature and ionic strength of the washing buffer.

TE [10 mM Tris–1 mM EDTA (pH 7.5–8.0)]   A buffer used in molecular biology for operations, such as dialyzing or solubilizing DNA, where only a dilute buffer is required.

tracking dye   A solution that is mixed with samples to be loaded onto a gel; it contains an agent, such as glycerol or Ficoll, to sink the sample into the well and dyes, such as bromphenol blue, which migrate at a known rate so that the migration of electrophoresed samples can be followed.

transformation   The process by which microorganisms accept and incorporate exogenously added DNA. Transformation occurs naturally among microbes or can be artificially induced using competent cells.

Tris [tris(hydroxymethyl)aminomethane]   An organic buffer used in molecular biology, biochemistry, and cell biology.

Triton X-100   A nonionic detergent.

unit of activity   Generally, the amount of an enzyme that catalyzes the conversion of a certain quantity of substrate to product in a fixed time period. For a restriction enzyme, 1 unit is defined as the amount required to digest 1 μg of standard DNA, such as phage λ, in 1 hour at optimum temperature and buffer conditions in a specified volume.

URA3   A gene used as a selectable marker for yeast.

YEPD (yeast extract–peptone–dextrose)   A complex medium for culturing yeast.

YNB (yeast nitrogen base)   A minimal medium for culturing yeast.

Index